MEDIEVAL MAPS OF THE HOLY LAND

MEDIEVAL MAPS OF THE HOLY LAND

P. D. A. HARVEY

London
THE BRITISH LIBRARY
2012

First published 2012 by
The British Library
96 Euston Road
London NW1 2DB

British Library
Cataloguing in Publication Data
A CIP record for this volume is
available from The British Library

ISBN 978 0 7123 5824 8

Designed by John Trevitt
Typeset in Poliphilus by
Norman Tilley Graphics Ltd,
Northampton
Printed in Hong Kong by
Great Wall Printing Co. Ltd.

CONTENTS

ILLUSTRATIONS

PREFACE

IT IS NOW MORE THAN FIFTEEN YEARS since I first appreciated how interesting our few medieval maps of the Holy Land are – and how much they contribute to our understanding of the way maps developed in the Middle Ages. Since then I have published one article on those by Matthew Paris and have given conference papers on several of the others, but I came to realise that they have enough in common to be treated as a group and discussed together. My first plan then was ambitious; this was to work towards a definitive edition that would cover not only the regional maps drawn in Europe but also the Islamic and Hebrew maps and the plans of individual cities – Jerusalem, Acre and Antioch. But it soon became clear that more ground-breaking work was needed before this could become remotely possible – we simply know too little of too many of the maps – and the present book is no more than a staging-post along the road that leads to that definitive edition.

It looks, and not in exhaustive detail, only at the eight regional maps in the Christian tradition that originated between the twelfth century and the fourteenth. The reader may well detect a certain emphasis on the maps as physical objects: how they were made, what changes they have undergone, how those in books relate to other material in the same volume and so on. This has come about naturally, not as deliberate policy; the answers to these questions are basic to our understanding of these maps, and they are questions that earlier writers on the maps have often overlooked.

In the course of work done intermittently over a decade and a half I have received much help from curators and other scholars, and this is acknowledged in many footnotes. To Dr Catherine Delano Smith, Professor Patrick Gautier Dalché and Professor Denys Pringle I owe an especial debt; beyond their help, specifically acknowledged, on particular points, they have allowed me to draw on their knowledge and experience of more general aspects of the work. I would also add a special word of thanks to Professor Jonathan Riley-Smith, the other leaders and organisers and all my companions on the 2010 St John Tour of the Levant that took me to the parts of the Near East on which the maps are focused, parts that I now saw for the first time; it contributed much to my understanding of the topography of the area, and this in turn must be reflected in what I have written. I – and all readers – owe a further debt to Professor Riley-Smith who, having read the text in draft, provided invaluable notes of points that needed correction and other comments.

For all they have done to bring the book to publication I am most grateful to David Way, Sally Nicholls, Norman Tilley and John Trevitt; it has been a great pleasure to work with them, as also with Michele Allan and Chris Orton of the Cartographic Unit, University of Durham. My wife, Yvonne, has been, as always, a constant help at every stage of the work; sadly, though, she was unable to join the tour of the Levant that she would have so much enjoyed.

May 2012 P.D.A.H.

ABBREVIATIONS

Adamnan, ed. Meehan: *Adamnan's De locis sanctis*, ed. by Denis Meehan, Scriptores latini Hiberniae 3 (Dublin, 1958)

Adams, *Seventh annual report*: F.B. Adams, *Seventh annual report to the Fellows of the Pierpont Morgan Library* (New York, 1957), pp. 14-17

Alexander and Temple, *Illuminated manuscripts*: J.J.G. Alexander and Elźbieta Temple, *Illuminated manuscripts in Oxford college libraries, the university archives and the Taylor Institution*(Oxford, 1985)

Baumgärtner, 'Das Heilige Land': Ingrid Baumgärtner, 'Das Heilige Land kartieren und beherrschen', in *Herrschaft verorten: politische Kartographie des Mittelalters und der Frühen Neuzeit*, ed. by Ingrid Baumgärtner and Martina Stercken (Zurich, 2012)

Beazley, 'New light': C. Raymond Beazley, 'New light on some mediaeval maps', *Geographical Journal*, 15 (1900), pp. 130-41, 378-89

Beazley, *Dawn*: C. Raymond Beazley, *Dawn of modern geography* (3 vols, London, 1897-1906)

Bede, 'De locis sanctis': Bede, 'De locis sanctis', ed. by J. Fraipont, in *Itineraria et alia geographica*, Corpus christianorum, series latina 175 (Turnhout, 1965), pp. 245-80

Bevan and Phillott, *Mediaeval geography*: W.L. Bevan and H.W. Phillott, *Mediaeval geography: an essay in illustration of the Hereford mappa mundi* (London and Hereford, 1873)

BL: London, British Library

Burchardus, Liber de descriptione: *Burchardus de Monte Sion, Liber de descriptione Terre Sancte: textus conferendus*, Société de l'Orient latin (privately printed, 1880)

Capello, *Mappamondo di Vercelli*: C.F. Capello, *Il mappamondo medioevale di Vercelli (1191-1218?)*, Università di Torino, Memorie e Studi Geografici 10 (Turin, 1976)

Cardini, 'Per un' edizione critica': Franco Cardini, 'Per un' edizione critica del *Liber secretorum fidelium crucis* di Marin Sanudo il vecchio', *Ricerche Storiche*, 6 (1976), pp. 191-250

Catalogue of manuscript maps, British Museum: *Catalogue of the manuscript maps, charts, and plans, and of the topographical drawings in the British Museum* (3 vols, London, 1844-61)

Connolly, 'Imagined pilgrimage': Daniel K. Connolly, 'Imagined pilgrimage in the itinerary maps of Matthew Paris , *Art Bulletin*, 81 (1999), pp. 598-622

Connolly, *Maps of Matthew Paris*: Daniel K. Connolly, *The maps of Matthew Paris: medieval journeys through space, time and liturgy* (Woodbridge, 2009)

Coxe, *Catalogus codicum*: H.O. Coxe, *Catalogus codicum qui in collegiis aulisque Oxoniensibus hodie adservantur* (2 vols, Oxford, 1852)

Cristoforo Colombo, ed. Cavallo: *Cristoforo Colombo e l'apertura degli spazi: mostra storico-cartografica*, ed. by Guglielmo Cavallo (Rome, 1992)

Degenhart and Schmitt, 'Sanudo und Veneto': Bernhard Degenhart and Annegrit Schmitt, 'Marino Sanudo und Paolino Veneto: zwei Literaten des 14. Jahrhunderts in ihrer Wirkung auf Buchillustrierung und Kartographie in Venedig, Avignon und Neapel', *Römisches Jahrbuch für Kunstgeschichte*, 14 (1973), pp. 1-137

Descriptiones, ed. Tobler: *Descriptiones Terrae Sanctae ex saeculo VIII. IX. XII. et XV.*, ed. by Titus Tobler (Leipzig, 1874)

Desimoni, 'Carta della Terra Santa': Cornelio Desimoni, 'Una carta della Terra Santa del secolo XIV', *Archivio Storico Italiano*, 5th series, 11 (1893), pp. 241-58

Edson, 'Matthew Paris' "other" map': Evelyn Edson, 'Matthew Paris' "other" map of Palestine', *The Map Collector*, 66 (Spring 1994), pp. 18-22

Four maps of Great Britain, ed. Gilson: *Four maps of Great Britain designed by Matthew Paris about 1250*, ed. by J.P. Gilson (London, 1928)

Gautier Dalché, 'Cartes de Terre Sainte': Patrick Gautier Dalché, 'Cartes de Terre Sainte, cartes de pèlerins', in *Fra Roma e Gerusalemme nel medioevo*, ed. by Massimo Oldoni (3 vols, Salerno, 2005), II, pp. 573-612

Gautier Dalché, 'Cartes, réflexion stratégique et projets': Patrick Gautier Dalché, 'Cartes, réflexion stratégique et projets de croisade à la fin du XIIIe et au début du XIVe siècle: une initiative franciscaine?', *Francia*, 37 (2010), pp. 77-95

Gesta dei per Francos, ed. Bongars: *Gesta dei per Francos*, ed. by Jacques Bongars (2 vols, Hanover, 1611)

Ghinato, *Paolino da Venezia*: P. Allarto Ghinato, *Fr. Paolino da Venezia O.F.M. vescovo di Pozzuoli (†1344)*, Studi e Testi Francescani 1 (Rome, 1951)

Gransden, *Historical writing*: Antonia Gransden, *Historical writing in England c.550 to c.1307* (London, 1974)

Guillaume de Tyr, ed. Huygens: *Guillaume de Tyr: chroniques*, ed. by R.B.C. Huygens, Corpus christianorum continuatio medievalis 63, 63A (2 vols, Turnhout, 1986)

Harvey, *Hereford map: introduction*: P.D.A. Harvey, *The Hereford world map: introduction* (London and Hereford, 2010)

Harvey, *Medieval maps*: P.D.A. Harvey, *Medieval maps* (London, 1991).

Harvey, 'Paris's maps of Britain': P.D.A. Harvey, 'Matthew Paris's maps of Britain', in *Thirteenth century England: IV*, ed. by P.R. Coss and S.D. Lloyd (Woodbridge, 1992), pp. 109-21

Harvey, 'Paris's maps of Palestine': P.D.A. Harvey, 'Matthew Paris's maps of Palestine', *Thirteenth century England: VIII*, ed. by Michael Prestwich, Richard Britnell and Robin Frame (Woodbridge, 2001), pp. 165-77

Hereford world map, ed. Harvey: *The Hereford world map: medieval world maps and their context*, ed. by P.D.A. Harvey (London, 2006)

Heydenreich, 'Jerusalem-Plan': Ludwig H. Heydenreich, 'Ein Jerusalem-Plan aus der Zeit der Kreuzfahrer', in *Miscellanea pro arte: Hermann Schnitzler zur Vollendung des 60. Lebensjahres am 13 Januar 1965*, ed. by J. Hoster and P. Bloch,

Schriften des Pro Arte Medii Aevi 1 (Düsseldorf, 1965), pp. 83-90, plates 1, 2

History of cartography: *The history of cartography*, ed. by J.B. Harley, D. Woodward et al. (6 vols, Chicago), i: *Cartography in prehistoric, ancient, and medieval Europe and the Mediterranean* (1987)

Imagining Jerusalem, ed. Donkin and Vorholt: *Imagining Jerusalem in the medieval West*, ed. by Lucy Donkin and Hanna Vorholt (Woodbridge, 2012)

Itinera, ed. De Sandoli: *Itinera hierosolymitana crucesignatorum (saec. XII-XIII)*, ed. by Sabino de Sandoli, Pubblicazioni dello Studium Biblicum Franciscanum 24 (4 vols, Jerusalem, 1978-84)

Itinéraires, ed. Michelant and Raynaud: *Itinéraires à Jérusalem et descriptions de la Terre Sainte*, ed. by Henri Michelant and Gaston Raynaud, Publications de la Société de l'Orient latin, série géographique 3 (Geneva, 1882)

Itineraria, ed. Cuntz: *Itineraria Antonini Augusti et Burdigalense*, ed. by Otto Cuntz, Itineraria romana 1 (Leipzig, 1929)

Itineraries of William Wey: *The itineraries of William Wey*, ed. by [George Williams], Roxburghe Club (London, 1857)

Jacoby, 'Crusader Acre': D. Jacoby, 'Crusader Acre in the thirteenth century: urban layout and topography', *Studi Medievali*, 3rd series, 20 (1979), pp. 1-45

James, *Manuscripts in Corpus Christi College, Cambridge*: M.R. James, *A descriptive catalogue of the manuscripts in the library of Corpus Christi College, Cambridge* (2 vols, Cambridge, 1912)

Kamal, *Monumenta cartographica*: Youssouf Kamal, *Monumenta cartographica Africae et Aegypti* (5 vols, privately printed, 1926-51)

Kretschmer, 'Marino Sanudo': Konrad Kretschmer, 'Marino Sanudo der ältere und die Karten des Petrus Vesconte', *Zeitschrift der Gesellschaft für Erdkunde zu Berlin*, 26 (1891), pp. 352-70, plates 8, 9

Kugler, *Ebstorfer Weltkarte*: Hartmut Kugler, *Die Ebstorfer Weltkarte* (2 vols, Berlin, 2007)

Leopold, 'Crusading proposals': A.R. Leopold, 'Crusading proposals in the fourteenth and fifteenth centuries', in *The Holy Land, holy lands and Christian history*, ed. by R.N. Swanson, Ecclesiastical History Society, Studies in Church History 36 (Woodbridge, 2000), pp. 216-27

Levi, *Itineraria picta*: Annalina and Mario Levi, *Itineraria picta: contributo alla studio della Tabula Peutingeriana*, Studi e Materiali del Museo dell'Impero Romano 7 (Rome, 1967)

Lewis, *Art of Matthew Paris*: Suzanne Lewis, *The art of Matthew Paris in the Chronica majora*(Aldershot, 1987)

Map book: The map book, ed. by Peter Barber (London, 2005)

Map of the Holy Land: *Map of the Holy Land illustrating the itineraries of William Wey, Fellow of Eton in A.D. 1458 and 1462. In fac-simile from the original in the Bodleian Library*, Roxburghe Club (London, 1867)

Matthaei Parisiensis chronica majora, ed. Luard: *Matthaei Parisiensis monachi Sancti Albani chronica majora*, ed. by Henry Richards Luard, Rolls Series 57 (7 vols, London, 1872-83)

Matthaei Parisiensis historia anglorum, ed. Madden: *Matthaei Parisiensis, monachi Sancti*

Albani, historia anglorum, ed. by Frederic Madden, Rolls Series 44 (3 vols, London, 1866–9)

Miller, *Mappaemundi* : Konrad Miller, *Mappaemundi: die ältesten Weltkarten* (6 vols, Stuttgart, 1895–8)

Nebenzahl, *Maps of the Holy Land*: Kenneth Nebenzahl, *Maps of the Holy Land: images of Terra Sancta through two millennia* (New York, 1986)

Nouv. biogr. gén.: *Nouvelle biographie générale*, ed. by J.C.F. Hoefer (46 vols, Paris, 1852–66)

ODNB: *Oxford dictionary of national biography*, ed. by H.G.C. Matthew and Brian Harrison (60 vols, Oxford, 2004)

Peregrinationes, ed. Huygens: *Peregrinationes tres: Saewulf, John of Würzburg, Theodericus*, ed. by R.B.C. Huygens, Corpus christianorum continuatio medievalis 139 (Turnhout, 1994)

Peregrinatores, ed. Laurent: *Peregrinatores medii aevi quatuor*, ed. by J.C.M. Laurent (2nd edn, Leipzig, 1873)

Pringle, *Churches*: Denys Pringle, *The churches of the crusader kingdom of Jerusalem: a corpus* (4 vols, Cambridge, 1993–2009)

Pringle, *Pilgrimage*: Denys Pringle, *Pilgrimage to Jerusalem and the Holy Land, 1187–1291*, Crusade Texts in Translation 23 (Aldershot, 2011)

Pujades i Bataller, *Cartes portolanes*: Ramon J. Pujades i Bataller, *Les cartes portolanes: la representació medieval d'una mar solcada* (Barcelona, 2007)

Röhricht, 'Karten und Pläne I': Reinhold Röhricht, 'Karten und Pläne zur Palästinakunde aus dem 7. bis 16. Jahrhundert, I', *Zeitschrift des Deutschen Palästina-Vereins*, 14 (1891), pp. 8–11, plate I

Röhricht, 'Karten und Pläne VI': Reinhold Röhricht, 'Karten und Pläne zur Palästinakunde aus dem 7. bis 16. Jahrhundert, VI', *Zeitschrift des Deutschen Palästina-Vereins*, 18 (1895), pp. 173–82, plates V–VII

Röhricht, 'Marino Sanudo': Reinhold Röhricht, 'Marino Sanudo sen. als Kartograph Palästinas', *Zeitschrift des Deutschen Palästinavereins*, 21 (1898), pp. 84–126, plates 2–11

Rorgo Fretellus, ed. Boeren: *Rorgo Fretellus de Nazareth et sa description de la Terre Sainte: histoire et édition de texte*, ed. by P.C. Boeren, Koninklijke Nederlandse Akademie van Wetenschapen, Afdeling Letterkunde, Verhandelingen, new series 105 (Amsterdam, 1980)

Sansone, *Mattheus Paris*: Salvatore Sansone, *Tra cartografia politica e immaginario figurativo: Matttheus Paris e l'Iter de Londinio in Terram Sanctam*, Istituto Storico Italiano per il Medio Evo, Nuovi Studi Storici 84 (Rome, 2009)

Simek, 'Hierusalem': Rudolf Simek, 'Hierusalem civitas famosissima', *Codices Manuscripti*, 12 (1992), pp. 121–53

Talbert, *Rome's world*: Richard J.A. Talbert, *Rome's world: the Peutinger map reconsidered* (Cambridge, 2010)

Tesi, *Monumenti*: Mario Tesi, *Monumenti di cartografia a Firenze (secc. X–XVII): catalogo* (Florence, 1981)

Thesaurus monumentorum, ed. Basnage: *Thesaurus monumentorum ecclesiasticorum et historicorum*, ed. by Jacques Basnage (4 vols, Antwerp, 1725)

Van de Walle, 'Deux fragments': Baudouin van de Walle, 'Deux fragments d'une carte médiévale de Palestine conservés aux Archives de l'état à Bruges', *Annales de la Société d'émulation de Bruges*, 71 (1928), pp. 56-75

Vaughan, 'Handwriting of Matthew Paris': R. Vaughan, 'The handwriting of Matthew Paris', *Transactions of the Cambridge Bibliographical Society*, 1, pt 5 (1953), pp. 376-94, plates XIV-XIX

Vaughan, *Matthew Paris*: Richard Vaughan, *Matthew Paris*, Cambridge Studies in Medieval Life and Thought, 2nd series 6 (Cambridge, 1958)

Vilnay, *Holy Land*: Zev Vilnay, *The Holy Land in old prints and maps*, transl. Esther Vilnay and Max Nurock (Jerusalem, 1963)

Warner and Gilson, *British Museum: catalogue of Royal manuscripts*: George F. Warner and Julius P. Gilson, *British Museum: catalogue of western manuscripts in the Old Royal and King's Collections* (4 vols, London, 1921)

Westrem, *Hereford map*: Scott Westrem, *The Hereford map: a transcription and translation of the legends with commentary*, Terrarum orbis 1 (Turnhout, 2001), reissued with the Folio Society's near-full-size reproduction, *The Hereford world map: mappa mundi* (London, 2010)

Wilkinson, *Pilgrimage 1099-1185*: John Wilkinson, *Jerusalem pilgrimage 1099-1185*, Hakluyt Society, 2nd series 167 (1988)

Wilkinson, *Pilgrims before the crusades*: John Wilkinson, *Jerusalem pilgrims before the crusades* (2nd edn, Warminster, 2002)

CONVENTIONS FOLLOWED IN THE TEXT

1. At the beginning of each of chapters 5-12 is a brief description of the map or maps that it covers. Conventions followed there are:

(i) Measurements are of the entire leaf or membrane or, if only part is used for the map, of the dedicated portion. They are given in centimetres to the nearest 0.5 cm, first vertical, then horizontal. Even the measurements that are not said to be approximate should be not taken as precise; none of the pieces of parchment is exactly rectangular.

(ii) By 'ink' is meant pigment applied with a pen, by 'paint' pigment applied with a brush. The colours are those of their present appearance.

(iii) By a 'glazed' frame is meant a covering that may be either glass or transparent plastic.

2. Place-names in italic type – e.g. Zoar (*Segor*) – show the spellings used on the map under discussion.

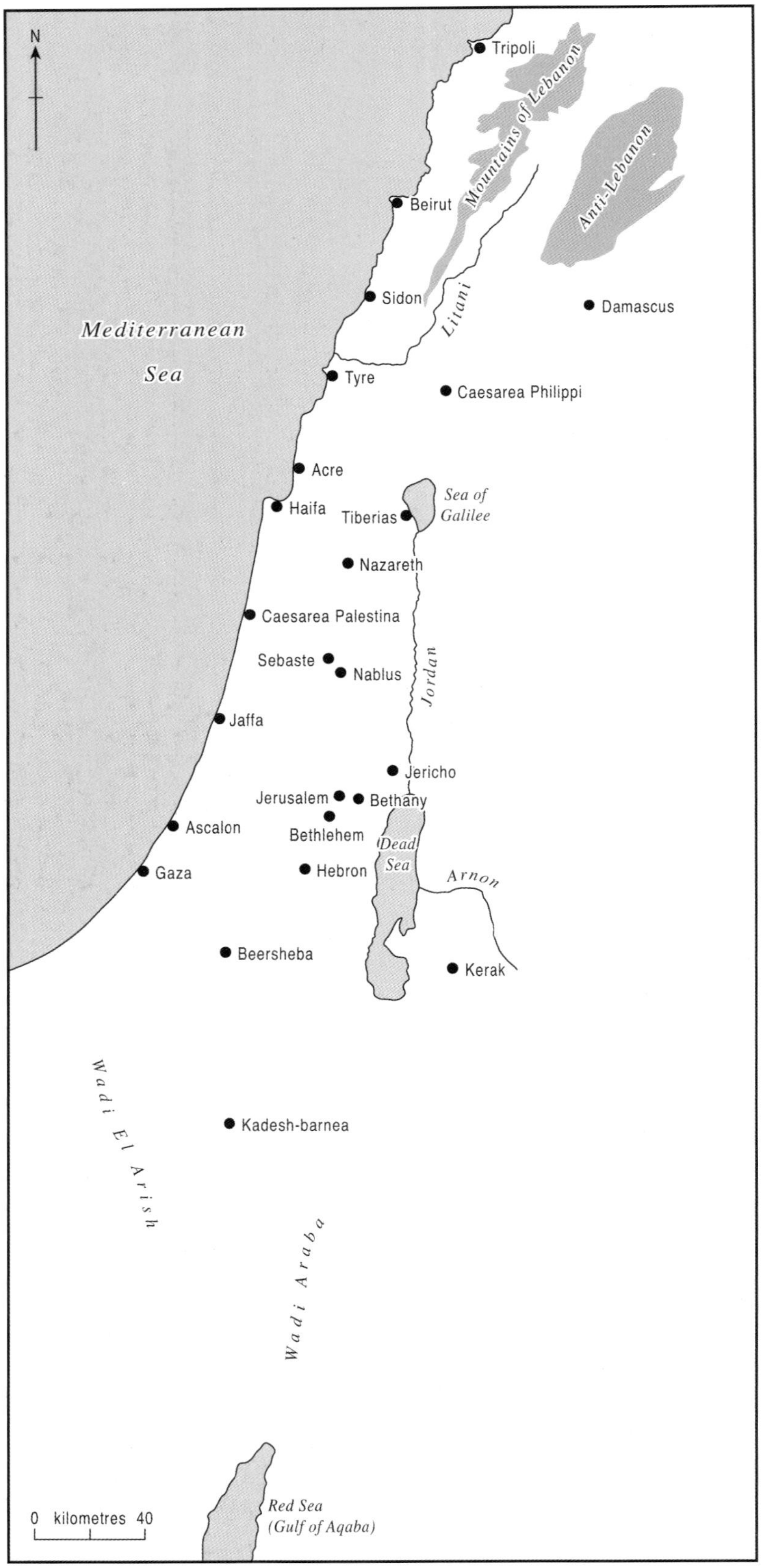

1. Map of Palestine showing the principal places mentioned in the book.

THE HISTORICAL BACKGROUND 1

THIS BOOK looks in detail at eight regional maps of Palestine that were first drawn between the late twelfth century and the mid-fourteenth; with their various versions and derivatives we know them through twenty-three surviving artefacts. They are the Ashburnham Libri map (chapter 5, p. 31), the three Tournai maps (chapter 6, p. 40), Matthew Paris's Oxford map (chapter 7, p. 60), Matthew Paris's Acre map (chapter 8, p. 74), the large Burchard map (chapters 9, p. 94, to 11, p. 128) and the small Burchard map (chapter 12, p. 141). In the Tournai maps and Matthew Paris's Acre map what we have is work of the original mapmaker, but of the others we have only copies and derivatives of original maps that are now lost. In the late nineteenth century interesting and detailed work was published on all these maps, but since then only Matthew Paris's maps and the grid-map abridgement of the large Burchard map have attracted attention. They are all, however, of considerable interest for the light they throw on the way maps were thought of and constructed in medieval Europe. They contribute too to our understanding of the way Palestine was viewed at a period when the crusades gave it particular interest, not only to chroniclers and historians down to the present day, but also to devout Christians throughout contemporary western Europe.[1]

'Palestine' in this book is used purely as a geographical term, with no historical or political implications whatsoever. It is to be understood as the area between the river Jordan and Wadi Araba on the east, the Mediterranean on the west, and stretching from the mountains of Lebanon in the north to Wadi El Arish in the south. But although all the maps discussed are focused on this area, only one (the first Tournai map) is confined to it, and they set Palestine in a geographical context that may extend to the Orontes in the north, the Euphrates in the east, the Nile in the south – and even well beyond.

Palestine, then, is a small area, about the same size as Wales. However, it lies across routes from Africa into Asia, and thus on the way between the early civilisations of Mesopotamia and the Nile, and this underlies its extraordinary importance for religious and cultural history, for Judaism and for the two other great monotheistic religions that have sprung from it. For Jews it is the land where Abraham settled, the Promised Land to which Moses led the Israelites from Egypt; it was here that David founded Jerusalem, the city where his son Solomon built the temple that lay at the heart of Jewish faith. For Muslims it is the land associated with Abraham, Moses and Jesus, precursors of the prophet Mohammed; Jerusalem is one of Islam's three holy cities, along with Mecca and Medina. And for Christians it is the Holy Land where Jesus was born and ministered, and Jerusalem the city where he achieved mankind's salvation through his death on the Cross.

1 Baumgärtner, 'Das Heilige Land', surveys these same maps, looking more closely than here at their underlying political implications. I am most grateful to Prof. Baumgärtner for letting me see this interesting article before publication.

All this, together with its geographical position, has contributed to the area's complicated political history. Over the past three thousand years – which is about as far back as we know – Palestine's longest period of relative stability was as part of the Ottoman empire, from the early sixteenth century to the early twentieth. Here, by way of introduction to the maps, we shall look only at some of the more striking developments from the first century AD to the eleventh, then, in broadest outline, at what happened between the last years of the eleventh century and the end of the thirteenth. This should not be seen as any kind of consistent narrative of events; it is meant simply to anticipate some questions that might arise on looking at the maps.

In the early first century the events of the Gospels were played out in two areas of the Roman empire, Judea to the south under the direct rule of a procurator, and Galilee to the north, where there were still vestiges of the semi-autonomy that the Jewish kingdom of Judea had enjoyed before the death of Herod the Great in 4 BC. Increasing tensions in the mid-first century led to the Jewish revolt that broke out in 64; it ended with the Romans' destruction of the temple at Jerusalem in 70 and the last resistance at Masada by the Dead Sea in 73. In a further rebellion in 132-5 Jerusalem itself was destroyed and was then rebuilt as a new Roman city, Aelia Capitolina. While many Jews stayed in Palestine – indeed, some communities remained in Galilee to the time of the crusades and later – many more left over these years to join the large groups already settled in Mesopotamia, Egypt and elsewhere. Palestine was no longer to be seen as a Jewish land.

In the division of the Roman empire between east and west, at first pragmatic as occasion demanded but formal and permanent from the end of the fourth century, Palestine was naturally in the Greek-speaking eastern empire. Meanwhile, Constantine, the first Christian emperor, had founded Constantinople as his new capital in the 320s. It came to be believed that his mother, Helena, had miraculously discovered at Jerusalem the cross on which Jesus had been crucified, and this can be seen in the context of a new interest in Palestine as the site of mankind's redemption. It was here that the places named in the Gospels were to be found and could be visited; Helena is credited with starting the work of building on the holy sites. The age of pilgrimage had begun.[2]

In the early seventh century Christian rule in Palestine was first interrupted, then ended. The Persian empire centred on the area between the Caspian Sea and the Persian Gulf had long had uneasy relations with the eastern emperor. From the middle of the century's second decade the Persians under Chosroes II occupied the lands around the eastern end of the Mediterranean – Palestine was taken in 614 – and held them until they were recovered by the emperor Heraclius in 627. Historians mark the reorganisation of the empire that followed by giving it a new name, the Byzantine empire, from the older name of the site of Constantinople. Ten years later, however, Palestine, with Egypt, Mesopotamia and much else, was claimed by the explosive expansion of Arab rule that followed the death of Mohammed in 632. Palestine was taken over after a decisive battle on the river Yarmuk, east of the sea of Galilee, in 634. The new empire, or caliphate, that soon stretched from Spain to India, came to be dominated by two successive dynasties, the Umayyads at Damascus from 661 to 749 and the Abbasids who set up a new capital at Baghdad. While there

2 Among accounts of these developments are E.D. Hunt, *Holy Land pilgrimage in the later Roman empire AD 312-460* (Oxford, 1982), and Robert L. Wilken, 'Eusebius and the Christian Holy Land', in *Eusebius, Christianity and Judaism*, ed. by Harold W. Attridge and Gohei Hata (Detroit, 1992), pp. 736-60, which discusses the identification of the holy places and the concept of a Holy Land.

was every inducement to convert, and many must have done so, Arab conquest did not mean that the entire population became Muslim; Jews and Christians were subjected to some disabilities, but the Koran enjoined toleration towards them and the succession of Christian patriarchs at Jerusalem continued uninterrupted until 969, when one was executed for treasonable correspondence with the Byzantines.

By the mid-eleventh century the empire of the caliphate had long broken up into separate states; Palestine was then under the rule of the Fatimid caliph in Egypt. However, a new power threatened both the Arab states and the Byzantine empire. The Turkish peoples east of the Caspian Sea had been converted to Islam in the tenth century; now, under the successors of their leader Seljuk, they began aggressive incursions into the lands at the eastern end of the Mediterranean. In 1071 a victory against the Byzantines at Manzikert, near Lake Van in what is now eastern Turkey, opened their way to occupying most of Asia Minor apart from the sea coasts. In the hope of help from western Europe to regain their losses, the emperor Alexius Comnenus sent an embassy to the church council of Pope Urban II at Piacenza in 1095, urging the sufferings of the Christians of Asia Minor under Turkish rule. This was one theme of another council the pope summoned later that year at Clermont, but by now it had become part of a wider ambition, the duty – the necessity even – of rescuing the holy places of Palestine from the hands of the infidels. The result was the first crusade.[3]

The crusaders, probably more than sixty thousand of them in contingents drawn from various parts of western Europe under separate leaders, gathered at Constantinople and in 1097 moved across Asia Minor towards the Holy Land. There the situation was less straightforward than it may have seemed in the west. Since the 1070s the Seljuk Turks and the Fatimid caliphs had been disputing possession of Syria and Palestine; Jerusalem had been taken by the Seljuks in 1071 but was retaken by the Fatimids in 1098. This struggle enormously helped the crusaders, as neither side could give the newcomers its undivided attention; whereas the crusaders saw all Muslims as the enemy, the Muslims saw the Christians invaders as no more than a diversion from the long-term struggle between themselves. The crusaders – *Franci* in contemporary Latin chronicles, Franks to historians – captured Jerusalem in their turn in 1099 and set up their own state, the kingdom of Jerusalem. Three other crusader states established to the north came to be nominally under the suzerainty of the kings of Jerusalem: the county of Edessa, the principality of Antioch and the county of Tripoli. The areas of these states were constantly shifting and in the long run were contracting, but at their greatest extent they included the whole of Palestine and the coastal areas north to the Orontes, thence extending east to the Euphrates and beyond.

We look in vain on our medieval maps for the boundaries – or even the existence – of these crusader states. What we do find on the maps, however, are the sites of many of the crusaders' castles and, indeed, castles of their Muslim enemies as well. The twelfth century was a period when many castles were built throughout western Europe, so it is unsurprising that the crusaders built them in the lands they held, where their position, if not always under threat, was at best precarious. They were built by the rulers, by nobles and by the great military orders of knights who lived and

3 A great deal has been written about the crusades. *A history of the crusades*, ed. by Kenneth M. Setton (6 vols, vol. 1 Philadelphia, vols 2-6 Madison, 1955-89), is a comprehensive multi-author work; Steven Runciman, *A history of the crusades* (3 vols, Cambridge, 1951-4), gives a shorter but blow-by-blow account, with full references to primary sources; Jonathan Riley-Smith, *The crusades: a history* (2nd edn, London, 2005) is both readable and detailed, with an up-to-date bibliography.

fought under religious vows; these were the Hospitallers and the Templars, whose organisations dated from the second decade of the twelfth century, and the Teutonic Knights, founded at the end of the century. The castles were built as strongholds, places that could be effectively defended and that would serve as bases for attack, but they were built also as centres of administration. Some were in towns, but many (and it is these that we see named on the maps) were isolated, often built on easily defended heights that offered wide views across the land. Many changed hands, some more than once; the castle at Kerak, east of the Dead Sea, was built in 1142 by a vassal of the king of Jerusalem as one of the kingdom's outmost defences but, lost by the Franks in 1188, was by the late thirteenth century being used as a treasury by the sultan of Egypt.[4]

From the mid-twelfth century the Franks had to contend successively with three great Muslim leaders. The first was Zengi, who in 1144 took Edessa, bringing to an end one of the crusader states. This gave rise to the second crusade, led by the kings Conrad III of Germany and Louis VII of France, which reached the Holy Land in 1148 but which collapsed after its failure to take Damascus. The second great leader was Zengi's son, Nur-ed-Din, but the greatest challenge came from Saladin, who from his base as ruler of Egypt established himself in Aleppo and Damascus and who saw the destruction of the crusader states as both a political and a religious duty. His victory at Hattin, west of the sea of Galilee, in 1187 gave him effective mastery of all Palestine and the lands to the north; he took Jerusalem and by the end of the following year the crusader states were more or less confined to a few places on the coast. The city of Tyre was all that was left of the Franks' kingdom of Jerusalem.

The immediate consequence was the third crusade, and we may note in passing that it was at about this time that the earliest of our eight regional maps were drawn. In 1190 a large army failed to reach the Holy Land as an effective force after the death of its leader, the emperor Frederick I Barbarossa, in Asia Minor, but in the following year armies under the kings of France and England, Philip II Augustus and Richard I *Coeur-de-Lion*, arrived and succeeded in consolidating and improving the Franks' position. Acre was taken back and a treaty of 1192 gave the kingdom of Jerusalem a coastal strip with its cities, from Jaffa in the south to Beirut in the north; beyond, Tripoli (between Beirut and Tortosa) and Antioch remained in Christian hands. Saladin died in 1193.

A fourth crusade in 1202-4 turned into a war against the eastern emperor at Constantinople and many of those who set out came no further east, while a fifth at first succeeded only in bringing Damietta, on the Nile delta, under Christian control from 1219 to 1221. However, the emperor Frederick II, grandson of Barbarossa, arriving late for the crusade in 1228, achieved the restoration of Jerusalem, more by diplomacy than by force of arms. By a ten-year treaty in 1229 the city was returned to the kingdom to which it had given its name, along with Bethlehem and a strip of land that formed a corridor to the coast at Jaffa, as well as, in Galilee, some inland areas between the Jordan and the coast. It was not long, however, before Jerusalem was finally lost. In 1244 a large force of Khorezmian Turks from central Asia, incited by Ayub, sultan of Egypt, to invade Palestine, took the city; the army of the Franks,

4 *Peregrinatores*, ed. Laurent, pp. 58-9.

joined in alliance with Ayub's Muslim enemies, advanced towards Egypt, but was heavily defeated at La Forbie, near Gaza.

There followed another disastrous crusade, led by Louis IX, Saint Louis, king of France. Arriving at Damietta in 1249, his army occupied this city without difficulty, but a battle at Mansourah, thirty miles to the south, in February 1250 was won only at great cost and on its way back to Damietta in April the army was attacked and destroyed and Louis himself was taken prisoner. He was released a month later for a large ransom and sailed to Acre, where he remained for four years, effectively as ruler of the Frankish kingdom. While he was imprisoned a revolution against Ayub's successor had brought the Mamluks, his generals, to power in Egypt, leading to renewed enmity between Egypt and Syria. However, in 1259 a new situation arose with the arrival in Syria of Mongol invaders who had already occupied Persia and Mesopotamia; by May 1260 they had taken Damascus and were threatening Egypt, but in September the Mamluks, having secured the acquiescence of the Franks at Acre, won a victory at Ain Jalut (the pool of Goliath) ten miles south of Mount Tabor, which ended the Mongol threat and once more united Syria with Egypt, under Mamluk rule.

After 1260 this meant the rule of the sultan Baibars. With more determination than earlier Muslim leaders he sought to reduce – or eliminate – the Franks' presence in the east. In 1266 the capture of the Templars' castle at Safed brought Galilee under his control, and in 1268 he took Jaffa; in the same year he took Antioch and by his death in 1277 little was left to the Franks beyond the coastal cities from Acre to Tortosa with Latakia and the castles of Margat and Chastel Pèlerin. The crusade of 1271-2 led by the king of England's son, about to succeed his father as Edward I, achieved nothing, and the successors of Baibars continued his work. Tripoli fell to the Mamluks in 1289 and finally in 1291 they took and destroyed Acre after a siege of more than six weeks. The remaining coastal towns were taken in the weeks following, bringing to an end the last remnants of the crusader states.

Contemporary chroniclers tell us more about the crusades than we know about any of Palestine's earlier regime changes – which we have only touched on here – and the picture they give is of appalling mayhem and slaughter. This should not mislead us. We have looked at the history of the period in black-and-white terms, which is how it appeared to the first crusaders and, over the next two centuries, to the Christians of western Europe. In reality the picture was much more complicated. There was much interaction – both commerce and amity – between the Frankish settlers and their Muslim neighbours, both rulers and ruled, a fact that some of the later crusaders from Europe found disconcerting. For those not directly involved, whether inhabitants or pilgrims, warfare was probably less disruptive than we might imagine: Denys Pringle points out that Thietmar's account of his travels in Palestine and Sinai in 1217-18 never refers to the fifth crusade that was then already in progress.[5] We may reasonably suppose that in earlier centuries too the turmoil and violent changes that our sources describe scarcely affected the lives of most of the region's inhabitants. As William Dalrymple has put it, 'In the Middle East, the reality of continuity has always been masked by a surface impression of cataclysm.'[6]

5 Pringle, *Pilgrimage*, p. 51.

6 William Dalrymple, *From the Holy Mountain* (London, 1997), p. 184.

2 THE LITERARY BACKGROUND

IN THE NEXT CHAPTER we shall look at the cartographic background of our regional maps of Palestine and at the other maps that were being produced in the same period. However, this is only part of the story, for all our maps are no less closely related to written texts. To start with, every one is directly or indirectly related to the Bible, if only in the choice of places that are marked and named. The scenes of the greatest events of the New Testament – Nazareth, Bethlehem, Jerusalem – all appear and others as well: Cana, Capernaum, Emmaus and so on. The narrative books of the Old Testament are less well represented on most of the maps, but we find Zoar (*Segor*), Endor, Shiloh and many others.[1] On the other hand it is unlikely that any of our mapmakers drew material directly from the Bible. Our two latest maps, the large and small Burchard maps, include a mass of biblical information – not only the names of places, but notes of what happened there – and this comes not from the Bible itself but from the account of the Holy Land by Burchard of Mount Sion. The names on the earlier maps will have been taken from the world maps that provided their basic outlines or from other descriptive writings; the biblical information reached the maps at second hand. All the same we should not lose sight of the fact that, one way or another, the Bible underlies everything these maps show us; the events of the Bible and their long-term consequences are what all these maps are about.

The Bible of our mapmakers and their sources was what came to be known as the Vulgate, the Latin translation by St Jerome authorised by the Church. Born in Dalmatia in about 340, Jerome was a learned and much travelled man; from 382 to 384 he was at Rome as secretary to Pope Damasus, at whose suggestion he began his translation, and from 386 until he died in 420 he was head of a religious house in Bethlehem. He produced many other works, among them commentaries on the books of the Bible, lives of holy men and writings on theological controversies. One of his works was a translation of the *Onomasticon* written in Greek by Eusebius, a gazetteer of places named in the Bible, in many cases giving the distance from one to another. His translation is known as *The names and locations of places of the Hebrews* (*De situ et nominibus locorum hebraicorum*) but is often given a shorter title, *The book of places* (*Liber de locis* or *Liber locorum*).[2]

We have accounts of Palestine, along with many other lands, from long before the time of Jerome. Pliny the elder, who died in AD 79, described the world's geography in a section of his vast *Natural history* (*Naturalis historia*); in Palestine he tells us, for instance, that the peninsula on which Tyre is built was once an island, 700 paces from the coast, and that the Dead Sea is over 100 miles long, 75 miles across at its widest point, 6 miles at its narrowest.[3] He draws on the work of earlier authors, and later

1 Zoar, Genesis 19.22, etc. (Ashburnham Libri, Matthew Paris Oxford, Large Burchard); Endor, 1 Samuel 28.7, etc. (Ashburnham Libri, Second and Third Tournai, Large Burchard); Shiloh, Joshua 18.1, etc. (Ashburnham Libri, Second and Third Tournai, Large Burchard).

2 Printed in *Patrologiae cursus completus, series latina*, ed. by J.-P. Migne, 23 (Paris, 1845), cols 859-928.

3 Pliny, *Natural history*, ed. by H. Rachman et al., Loeb Classical Library (10 vols, London and Cambridge Mass., 1945-80), II, pp. 274-5, 278-9.

authors drew on his: the geographical sections of the early-third-century *Collection of memorable matters* (*Collectanea rerum memorabilium*) by Gaius Julius Solinus and the encyclopaedic *Marriage of Philology and Mercury* (*De nuptiis Philologiae et Mercurii*) by Martianus Capella, who was a contemporary of Jerome, both took much from Pliny. Another contemporary of Jerome was Paulus Orosius, whose *Seven books of history against the pagans* (*Historiarum adversum paganos libri septem*) also included an account of the world's geography, and so too did the encyclopaedic works of Isidore of Seville in the seventh century and Lambert of Saint-Omer in the twelfth. If any of these works contributed at all to our regional maps of Palestine, they did so indirectly: they were used as sources of information for world maps in the twelfth and thirteenth centuries, the world maps that played a significant role in constructing most of our regional maps.[4]

Also from the Roman period we have two important itineraries, lists of successive places along routes with the distances between them. The so-called Antonine Itinerary, from the third century, is a functional reference work; it covers the entire Roman empire, route by route, giving the distance from one place to the next but adding no further information. Thus, on the way from Antioch to Alexandria it is 18 miles from Arqa to Tripoli, then a further 32 to Byblos, 24 to Beirut, 30 to Sidon, 24 to Tyre, 32 to Acre (under its older name *Ptolomaidam*) and so on.[5] The Bordeaux Itinerary, however, is a more personal record, the account of a journey to the Holy Land made by an anonymous pilgrim from Bordeaux in 333-4. Much is simply in itinerary form, giving the distances from place to place, distances that differ from those in the Antonine Itinerary; thus along the coast to Palestine our pilgrim notes further staging-posts and omits Byblos, but his distance from Arqa to Tripoli is 16 miles, then 48 to Beirut, 28 to Sidon, 25 to Tyre and 33 to Acre (*Ptolemaida*). At certain places he adds a brief note – Sarepta, for instance, was where Elijah asked the widow for food[6] – and from Nablus to Jerusalem and then on to Hebron the itinerary turns into a full description, especially of Jerusalem itself, with its sites of events in both the Old and the New Testaments.[7]

In this we can see the Bordeaux Itinerary as the precursor of an entire genre of writing that continued throughout the Middle Ages: accounts of the Holy Land, of varying length and varying degrees of detail. Some occupy no more than a page or so of a manuscript, others a substantial part of a volume. Most are focused on Jerusalem, with descriptions of the holy sites in and around the city, but most also describe Bethlehem, Nazareth and other places in Palestine that the pilgrim might visit, or simply pass through. Their interest, literary quality and historical value vary as much as their length and detail. We shall look more closely at the account by Burchard of Mount Sion when we look at the maps based on it (chapter 9, pp. 94-100); written in the late thirteenth century it is the longest and in many ways the most impressive of these works. By then a great many others had been written – they are to be found in many medieval manuscripts from all over Europe – and of course they continued to be written after the thirteenth century.

Many of the earliest of these writings, and some later ones, are personal travel accounts, describing the writers' own experiences as pilgrims in the Holy Land: where they went and what they saw. An early, unfortunately incomplete, example is

4 Thus their contributions to the Hereford world map are described by Westrem, *Hereford map*, pp. xxix-xxxiii.

5 *Itineraria*, ed. Cuntz, , p. 21.

6 1 Kings 17.10-16.

7 *Itineraria*, ed. Cuntz, pp. 94-8.

the account by Egeria from Gaul, who may or may not have been a nun and who spent three years in Palestine in the early 380s.[8] In at least one notable case we may wonder whether the travel account is not just a literary device. Adomnan, abbot of Iona, who died in 704, wrote down what was recounted to him by Arculf, an otherwise unknown Gallic bishop; his account of his travels includes plans of the church of the Holy Sepulchre and Golgotha, the basilica on Mount Sion, the church on the site of the Ascension and the church at Jacob's well at Sychar. Bede a little later wrote a revised version of Adomnan's text, including three of the four plans and explaining that when Arculf was returning to Gaul storms drove his ship on to the west coast of Britain, where he met Adomnan; even if this was not at Iona but further south, the explanation raises some sceptical doubts.[9] But even if this is not how the material reached Adomnan, it in no way detracts from the work's value as someone's description of the Holy Land in the seventh century, after the Arab conquest. Certainly there are many later undoubted personal accounts of travels to the Holy Land, among them those of the Englishman Saewulf in 1102-3[10] and – one we shall refer to in looking at the later large Burchard maps (chapter 11, pp. 135-6, 138-9) – of another Englishman, William Wey, in the mid-fifteenth century.

From the period of the crusades we have accounts of Jerusalem and the Holy Land by Frankish settlers, or included in chronicles of events. One notable description is by Rorgo Fretellus, who was archdeacon of Nazareth in the 1140s, while William of Tyre's chronicle includes a description of Jerusalem that Matthew Paris adapted in his.[11] But many short, anonymous texts also date from the twelfth and thirteenth centuries, in French as well as in Latin. Some of these texts form intricately connected groups; we shall see how the Ashburnham Libri map drew on a text from one such group in the twelfth century, Matthew Paris's Oxford map on one from another in the thirteenth (chapters 5, p. 39 and 7, p. 71).

Many of these descriptions of Palestine have been printed, but many have not; thus the account of places in the Holy Land that the monk who drew the Tournai maps copied in another volume has never been noticed in print (chapter 6, pp. 41, 52). Of those texts that have been printed, few have been definitively edited from all known manuscripts. We shall see how even for the work of Burchard of Mount Sion we have to use an inadequate edition (chapter 9, pp. 99-100) and for the no less important work of Marino Sanudo a text printed from a single, but by no means the only, manuscript in 1611 (chapter 10, pp. 107-9, 125). Many of the texts were printed in the late nineteenth century, carefully and with commentary that is learned and helpful – one thinks among others of those edited by W.A. Neumann and Titus Tobler – but some of these published texts are based on a single manuscript and make little attempt to explore the admittedly complex relationships between the texts. However, the collections of texts brought together in the same period by the Société de l'Orient latin and (mostly in English translation) by the Palestine Pilgrims' Text Society at least make it easy for readers to make their own comparisons, while Sabino de Sandoli in 1978-84 published a valuable collection of reprinted texts, with a few newly edited, from the period of the crusades.[12] De Sandoli accompanied the Latin and French texts with an Italian translation and there have been recent English translations of texts down to the twelfth century in two publications by John Wilkinson, with a

8 *Itinerarium Egeriae (Peregrinatio Aetheriae)*, ed. by Otto Prinz (Heidelberg, 1960); John Wilkinson, *Egeria's travels to the Holy Land* (2nd edn, Jerusalem and Warminster, 1981), provides an English translation.

9 *Adamnan*, ed. Meehan; Bede, 'De locis sanctis'. The role of Arculf is described in Meehan's edition of Adomnan , pp. 34-5, 46-7, 62-3, 68-9, 90-1, and in Fraipont's of Bede, pp. 279-80; a second account by Bede is in *Bede's Ecclesiastical history of the English people*, ed. by Bertram Colgrave and R.A.B. Mynors (Oxford, 1969), pp. 506-9. A recent discussion is by Thomas O'Loughlin, 'Adomnán and Arculf: the case of an expert witness', *Journal of Medieval Latin*, 7 (1997), pp. 127-46.

10 *Peregrinationes*, ed. Huygens, pp. 59-77.

11 *Rorgo Fretellus*, ed. Boeren; *Guillaume de Tyr*, ed. Huygens, I, pp. 381-4; *Matthaei Parisiensis chronica majora*, ed. Luard, II, pp. 107-9.

12 *Itinera*, ed. De Sandoli. In the following chapters Sandoli's work is cited for references to most of these texts.

helpful full gazetteer,[13] and of the later texts by Denys Pringle with valuable historical and topographical notes.[14] And, gradually, good modern editions of these texts are appearing, among them the work of Rorgo Fretellus by P.C. Boeren in 1980, the *Estoire d'Outremer*, with its descriptions of Galilee, by Margaret A. Jubb in 1990 and the texts of Saewulf, John of Würzburg and Theoderic by R.B.C. Huygens in 1994.[15] All the same, there is a long way to go before we can have full confidence in the published texts at our disposal; the light they might throw on the regional maps is only a small part of their potential importance.

13 Wilkinson, *Pilgrimage 1099–1185*; Wilkinson, *Pilgrims before the crusades*, with gazetteer on pp. 277–370.

14 Pringle, *Pilgrimage*.

15 *Rorgo Fretellus* ed. Boeren; *A critical edition of the Estoires d'Outremer et de la naissance Salahadin*, ed. by Margaret A. Jubb, Westfield Publications in Medieval Studies 4 (London, 1990); *Peregrinationes*, ed. Huygens.

3 THE CARTOGRAPHIC BACKGROUND

MAPS were not a part of everyday life in medieval Europe.[1] Where now we would think naturally of drawing a map to set out a route or the boundaries of property or the streets of a town, what came naturally to people in the Middle Ages was to write a verbal description, even a long and complicated one. Apart from, as we shall see, maps of the world and, latterly, sea charts, the idea of representing land and what is on it in graphic form hardly ever occurred to anyone; when it did, we should see it as a strikingly original initiative. We see this clearly in local maps, maps drawn from the personal knowledge of the mapmaker. Very few exist; taking two areas where they have been systematically recorded, we have some three dozen from England, fewer than twenty from the Low Countries. Of these only four, all from England, date from the twelfth and thirteenth centuries; none of the others is earlier than the mid-fourteenth.[2] One might suppose that maps of small areas, perhaps roughly sketched for some immediate purpose, were often drawn but nearly always thrown away, or lost in the course of time. But other types of record, just as ephemeral, that relate, like maps, to land and landed property, were carefully preserved and survive in great profusion. The maps have not simply failed to survive; they were never drawn.

The Romans, however, did have maps and it has indeed been suggested that they used maps as extensively and for the same purposes as we use maps today, in laying out fields and towns, in planning military strategy, in administering provinces and so on. The evidence for this has been seriously questioned and it is no longer a generally held view, but there seems no doubt that the Romans drew and used maps a good deal more than people in the Middle Ages. They even drew plans to a consistent scale, a concept wholly unknown for much of the medieval period: from the first three centuries of the Christian era we have scale-plans of fields, of individual buildings and even of the whole of Rome, carved in stone as a permanent record. Probably more usual were picture maps, showing an area viewed as from a height, with features drawn not in plan but pictorially; these are the maps that we see, in later copies, illustrating treatises on surveying.[3]

And the Romans had maps of their world, probably what we would regard as reasonably good ones. This has been doubted, but when we look at medieval world maps we see on one or another the occasional outline that looks familiar: on the early-eleventh-century Cotton map the British Isles, the Frisian coast and the Peloponnese all bear some relation to their shape on a modern map **[2]** and on another eleventh-century map in a copy of the work of Macrobius we see the north-west to south-east alignment of Italy and the Adriatic and recognisable outlines of Corsica and Sardinia.[4] The eleventh century had neither the technical nor the administrative capacity to map large areas, nor, indeed, the slightest reason to do so. These relatively

1 There is a general account of medieval maps in *History of cartography*, pp. 283-501, and a short but well illustrated overview in Harvey, *Medieval maps*.

2 There is an account of medieval local maps by P.D.A. Harvey in *History of cartography*, pp. 464-501; those from England and the Low Countries are described and illustrated in *Local maps and plans from medieval England*, ed. by R.A. Skelton and P.D.A. Harvey (Oxford, 1986), and Pieter van der Krogt, 'Lokale kaarten van Nederland uit de late Middeleeuwen', *Caert-Thresoor*, 27 (2008), pp. 29-42.

3 The view that many maps were made and used in the Roman period is set out by O.A.W. Dilke in his *Greek and Roman maps*(London, 1985) and in *History of cartography*, pp. 201-57, 276-9. The evidence for this is questioned in detail by Kai Brodersen, *Terra cognita: Studien zur römischen Raumerfassung*, Spudasmata 59 (Hildesheim, 1995).

4 Munich, Bayerische Staats-bibliothek, Clm 6362, f. 74r, illustrated in Harvey, *Hereford map: introduction*, p. 46.

correct shapes can have reached medieval maps only – at many removes – from their ancestors in the Roman period, and when we put all these relatively correct shapes together we realise that these ancestral maps must have had outlines not so far removed from those we see on maps today.

At the same time, these relatively correct shapes on medieval world maps are few and far between. Most – nearly all – the outlines on these maps are unrecognisable travesties of geographic reality. This may have come about through changes made unconsciously by successive copyists; not knowing the true shapes and probably seeing them anyway as unimportant, they will have been as likely to exaggerate any errors as to correct them, and will have distorted outlines to fit the page of the book or the sheet of parchment they were working on. An interesting suggestion recently put forward is that shapes were deliberately altered to give symbolic meaning to the positions of key cities: they were placed so that they lay on the equilateral triangle of the Trinity or on concentric circles that represented perfection. The map should not just present meaningless geographical outlines; rather, it should be a statement of the divine ordering of the world, of the harmony of space and time.[5] Indeed, we see some such process at work on a map of Europe by Lambert of Saint-Omer in the early twelfth century; with one broad peninsula and four narrower ones it can be seen as the hand of God raised in benediction.[6] On the Hereford world map of about 1300 the north coast of the Mediterranean has been given the same shape as the body of the crucified Christ pictured at Jerusalem, a reflection of God's presence in the world.[7]

Simple diagrams of the world were often drawn in medieval Europe: either a T within a circle, the T-O map showing the inhabited world divided between the three continents, Asia, Europe and Africa, or a zonal map showing the earth from the side divided (usually) into five climatic zones, one torrid, two temperate and two frigid. In the Middle Ages it was known that the earth is a sphere, but it was thought to be smaller than it really is and it was understood that the torrid zone around the equator was so hot as to be impassable; mankind was confined to the northern hemisphere and what was to be found in the southern was a subject of learned speculation. The equatorial belt was sea, not land, so the southern shore of Africa lay north of the equator and the inhabited world could be represented, like a kind of projection, as a circle of land surrounded by ocean with a few off-shore islands, Britain and Ireland among them.

Occasionally one of these diagram maps might be expanded by adding a few geographical features, but until the fourteenth century we have very few more detailed productions. The earliest, from the eighth century, are the Albi map, showing the inhabited world like a horseshoe around the Mediterranean, and the rather more detailed Vatican map, a circle with the ocean islands.[8] Also in the eighth century Beatus of Liébana included a world map among the many brightly coloured illustrations to his commentary on the Apocalypse, though we know it only from rather varied copies that date from the tenth century to the thirteenth. These so-called Beatus maps included a picture of Adam and Eve in the earthly paradise and some copies show the twelve apostles in their mission to bring Christianity to the world.[9] We may see here the first germs of a genre of encyclopaedic world maps that developed in northern France and England, maps that set within their geographical framework a

5 Brigitte Englisch, *Ordo orbis terrae: die Weltsicht in den Mappae mundi des frühen und hohen Mittelalters*, Orbis medievalis 3 (Berlin, 2002), pp. 123-244, 568-87 and passim; *Map book* , pp. 38-9.

6 Hartmut Kugler, 'Europa pars quarta: der Teil und das Ganze im "Liber floridus"', in *Europa im Weltbild des Mittelalters: kartographische Konzepte*, ed. by Ingrid Baumgärtner and Hartmut Kugler, Orbis medievalis 10 (Berlin, 2008), pp. 55-9.

7 Massimo Rossi, 'The Hereford mappamundi: visibile parlare', in *Hereford world map*, ed. Harvey, pp. 185-9.

8 Albi, Bibliothèque municipale, ms 29, f. 57v, and Vatican City, Biblioteca Apostolica Vaticana, ms Vat.lat. 6018, ff. 64v-65r, both illustrated in *Map book*, pp. 39, 43.

9 John Williams, 'Isidore, Orosius and the Beatus map', *Imago Mundi*, 49 (1997), pp. 7-32; Peter Barber, 'Medieval maps of the world', in *Hereford world map*, ed. Harvey, pp. 1-4.

great variety of information, with pictures, drawn from many sources: the Bible, the apocryphal accounts of Alexander the Great, bestiaries that described strange animals and birds, geographical works that described strange peoples in distant lands and so on.

Our earliest maps of this genre were drawn in books. On the Cotton or Anglo-Saxon map, drawn probably at Canterbury in the early eleventh century **[2]**, are a few short notes describing regions and peoples, and in the extreme north-east a picture of a lion: 'Here lions abound' (*Hic abundant leones*). On the map in a twelfth-century manuscript now at Munich **[3]** the lion is in Africa, but we see other beasts too, as well as Noah's ark, Alexander's camp and altars, the desert monasteries of Egypt and a few geographical notes. The Sawley map, drawn in about 1200 probably at Durham, again supplies odd bits of miscellaneous information among its coastal outlines, rivers and cities; we see, for instance, the lands where the dog-headed people (*cynocephales*) and the cannibals (*antropophagi*) live, the earthly paradise with its four rivers and the region of Gog and Magog, shut in by Alexander beyond the Caspian Sea. Rather later, in the 1260s, a world map was drawn in a psalter, probably at London **[4]**; only 9 cm across, it has long been recognised as a tiny picture of a much larger map, and this is certainly how we should also see the Sawley map and probably the Munich map as well. They are scaled-down versions of maps that were drawn not in books but on large sheets of parchment or cloth.[10]

These four world maps are probably a fair sample of those that were produced in books from the eleventh century to the thirteenth; there can never have been many of

10 All four are illustrated and discussed by Peter Barber, 'Medieval maps of the world', in *Hereford world map*, ed. Harvey, pp. 4-13, 15-19.

2. Cotton world map, 1025 × 1050; east at the top. In the top left corner is a lion, 'Hic abundant leones' ('Here lions abound'). BL, Cotton MS Tiberius B.v, f. 56v.

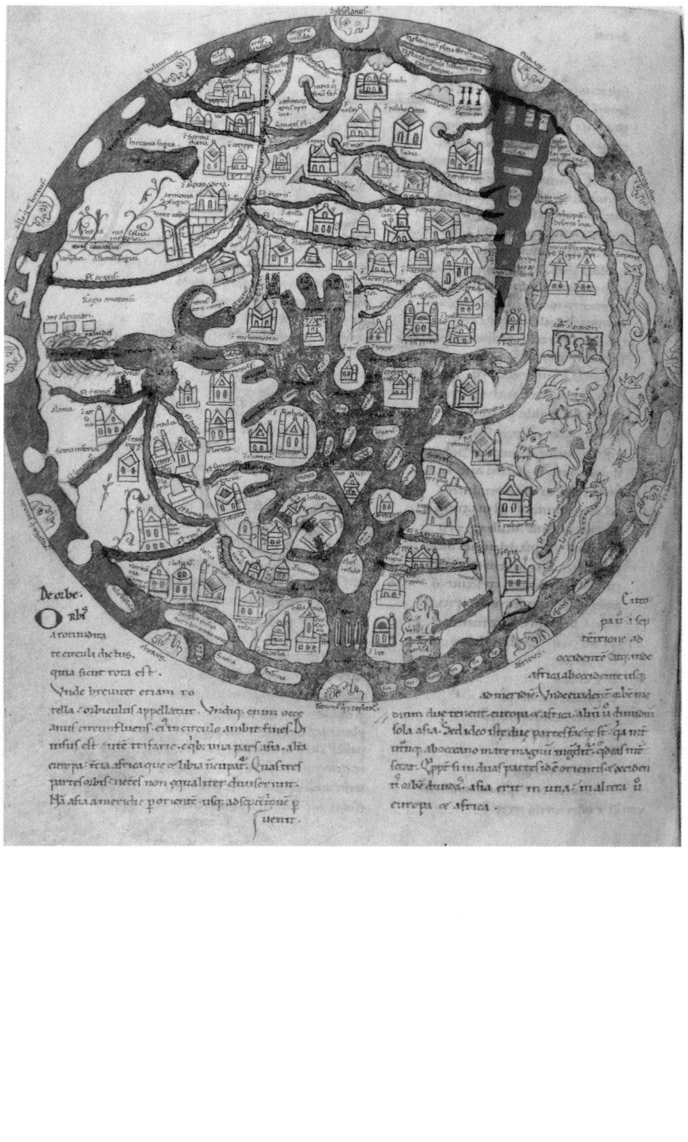

3. Munich world map, about 1130; east at the top. In Africa, along the right-hand side, are a lion and other beasts and to the right of the triangular Red Sea buildings mark the desert monasteries of Egypt and Alexander's camp. Munich, Bayerische Staatsbibliothek, Clm 10058, f. 154v.

them for, although many medieval books have been lost in the course of time, a great many survive that we might expect to contain maps of this kind. The encylopaedic world maps of this period were essentially a genre of large sheet maps, produced for display, for instruction and for contemplation of the wonders of God's world.[11] Certainly, as we shall see, the survivors of these large sheet maps are even fewer than the maps in books. But large sheet maps of any period – and this is a point we shall return to several times – are highly vulnerable; they become faded and discoloured if they are regularly displayed, they are clumsy and easily damaged and are willingly discarded as scrap once they appear old-fashioned or simply dirty and torn. It has been argued from internal evidence that these large world maps were produced in significant numbers, not perhaps hundreds but probably many dozens; their complex mass of detail will have placed severe demands on artist and scribe, making their construction slow and expensive.[12]

Of these large world maps there is only one complete survivor in finished form, the map of about 1300 that was made for Hereford Cathedral, still its home **[5]**. Drawn on a single sheet of parchment, 1.6 metres from top to bottom, 1.3 metres across, it has Jerusalem at the centre and outside the circle of the world Christ sits in judgment at the top, while in the two lower corners are enigmatic scenes that have been variously interpreted to explain why the map was made.[13] About four times the size of the

11 Patrick Gautier Dalché, 'Pour une histoire des rapports entre contemplation et cartographie au Moyen Âge', in *Les méditations cosmographiques à la Renaissance*, Cahiers V.L. Saulnier 26 (Paris, 2009), pp. 32-5.

12 P.D.A. Harvey, 'The Holy Land on medieval world maps', in *Hereford world map*, ed. Harvey, pp. 243-51.

13 Fully edited in Westrem, *Hereford map*; shorter accounts are Harvey, *Hereford map: introduction*, and Peter Barber, 'Medieval maps of the world', in *Hereford world map*, ed. Harvey, pp. 27-30.

4. Psalter world map, about 1265; east at the top. Only 9 cm in diameter, this is probably a miniature picture of a large wall-map. BL, Add. MS 28681, f. 9r.

5. Hereford world map, about 1300; east at the top. Our only complete example of the genre of encyclopaedic world maps that developed in England and northern France in the 12th and 13th centuries. Hereford, Mappa Mundi Trust.

Hereford map, some 3 metres square, was a map made for the convent at Ebstorf in Lower Saxony, probably a late-thirteenth-century copy of a map of about 1239 **[6]**. It was destroyed in an air raid on Hanover in 1943, but all available evidence has been used to create a careful reconstruction. It was drawn on thirty pieces of parchment and like the Hereford map was centred on Jerusalem, but the areas outside the map's circle were filled with accounts of the world, of places in it and of birds and animals, while the head, hands and feet of Christ protruded from behind the map, showing him supporting or underlying the world.[14] Much larger than even the Ebstorf map was a map 6 metres across that was painted on the wall of the church at Chalivoy-Milon, 40 miles from Bourges, in the mid-twelfth century; it was destroyed in 1885 and we know of it only from two descriptions written when it was already decayed.[15]

In the cathedral library at Vercelli is an early-thirteenth-century world map, drawn probably at Paris, that is best seen as a draft or cartoon from which a finished map might be constructed.[16] A fragment – one corner – of a late-thirteenth-century map very like the Hereford map but somewhat larger, survives in the archives of the Duchy of Cornwall, where it was used as a binding.[17] Two important contemporary descriptions of world maps have been published, setting out what was written on them,[18] and besides these various relics we have a number of references to what were probably encyclopaedic world maps, in Westminster Palace, Winchester Castle, Lincoln and Durham Cathedrals and elsewhere.[19]

As far as we can tell, these maps conformed more or less to a common pattern. They were circular, with east at the top, and the circle of land was surrounded by ocean with neighbouring islands that included Britain and Sri Lanka. The Mediterranean extended from the bottom of the map nearly to the centre, dividing Europe from Africa, and at its east end the Black Sea and the river Don on the north and the Nile on the south marked the bounds of Asia. Jerusalem might or might not be at the map's exact centre, but Palestine always occupied the central portion. Further east was India and in the extreme east at the top of the map was the earthly paradise. The Nile flowed from east to west, making a right-angled turn to enter the Mediterranean, and parallel to it, but flowing from west to east, was the Upper Nile, apparently unconnected to the Nile but actually flowing into it underground. Mountains were coloured brown and water green or blue, except the Red Sea and Persian Gulf which were red. Europe was dominated by its towns, and the map's encyclopaedic elements were distributed over the rest of the world, generally speaking the further away from north-west Europe the more wondrous. Thus along the southern shore of Africa, which, we should remember, lay north of the equator, was a series of monstrous races: those unable to open their mouth who had to take nourishment through a straw, those without a head whose eyes, nose and mouth were in their chest and several more.

Though maps in this tradition continued to be made in fourteenth-century England, the development of the encyclopaedic world map had halted by the end of the thirteenth century and, broadly speaking, the initiative in new forms of cartography had passed from north-west Europe to Italy. Here a new sort of map had come into being, the portolan chart, of which some thirty survive from the fourteenth century **[7]**. Our earliest example, the so-called *Carte pisane*, dates from the end of

14 Reconstructed and fully edited in Kugler, *Ebstorfer Weltkarte*; shorter accounts are Birgit Hahn-Woernle, *Die Ebstorfer Weltkarte*(Ebstorf, [1988]), and Peter Barber, 'Medieval maps of the world', in *Hereford world map*, ed. Harvey, pp. 23-7.

15 M. Kupfer, 'The lost *mappamundi* at Chalivoy-Milon', *Speculum*, 66 (1991), pp. 540-71.

16 Capello, *Mappamondo di Vercelli*, gives a full account; there is a colour reproduction in Rouben Galichian, *Countries south of the Caucasus in medieval maps: Armenia, Georgia and Azerbaijan* (Yerevan and London, 2007), pp. 159-60, and it is also illustrated and discussed in *History of cartography*, pp. 306-9, and in Harvey, *Hereford map: introduction*, pp. 55-6.

17 Graham Haslam, 'The duchy of Cornwall map fragment', in *Géographie du monde au Moyen Âge et à la Renaissance*, ed. by Monique Pelletier, Comité des travaux historiques et scientifiques, Mémoires de la Section de géographie 15 (Paris, 1989), pp. 33-44; Peter Barber, 'Medieval maps of the world', in *Hereford world map*, ed. Harvey, pp. 19-23.

18 Patrick Gautier Dalché, *La 'Descriptio mappe mundi' de Hugues de Saint-Victor* (Paris, 1988); Patrick Gautier Dalché, *Du Yorkshire à l'Inde: une 'Géographie' urbaine et maritime de la fin du XIIe siècle (Roger de Howden?)*, Hautes études médiévales et modernes 89 (Geneva, 2005), especially pp. 49-82.

19 *The history of the king's works*, ed. R. Allen Brown et al. (6 vols, London, 1963-82), I, p. 497; II, p. 861; P.D.A. Harvey, 'The Sawley map and other world maps in twelfth-century England', *Imago Mundi*, 49 (1997), pp. 36-40.

the thirteenth century,[20] but the genre was probably already in existence a century earlier.[21] The portolan chart is a sea chart with detailed coastal outlines; many coastal towns are named, always on the landward side so as not to intrude on the sea space. Overall are the so-called rhumb lines, which radiate from points placed in a circle around the map and were used for navigation. North is at the top and the magnetic compass clearly played a large part in the maps' construction and use. They were essentially practical charts for use by seamen, though those we have are more likely to be library copies, never taken to sea.[22]

To us today portolan charts look like real maps in a way contemporary and earlier world maps do not; their coastal outlines are close to those familiar to us. Our earliest examples, perhaps all produced in Genoa or Venice, cover accurately only the Mediterranean and the Black Sea; they had probably originated in sketched outlines that were progressively corrected through practical experience, and after regular voyages had begun from Italy to the English Channel in the second decade of the fourteenth century the Atlantic coasts as far as East Anglia and Flanders achieved more accurate outlines and greater detail. This process of enlargement was to continue, and in the fifteenth century it extended down the west coast of Africa. Long before this, portolan charts were being made outside Italy – in Majorca before

20 Illustrated in *History of cartography*, plate 30, and Harvey, *Medieval maps*, pp. 40-1.

21 Patrick Gautier Dalché, *Carte marine et portulan au XIIe siècle*, Collection de l'École française de Rome 203 (Rome, 1995), especially pp.19-22.

22 There are general accounts of portolan charts by Tony Campbell in *History of cartography*, pp. 371-463, and by Pujades i Bataller, *Les cartes portolanes* (English translation on pp. 401-526).

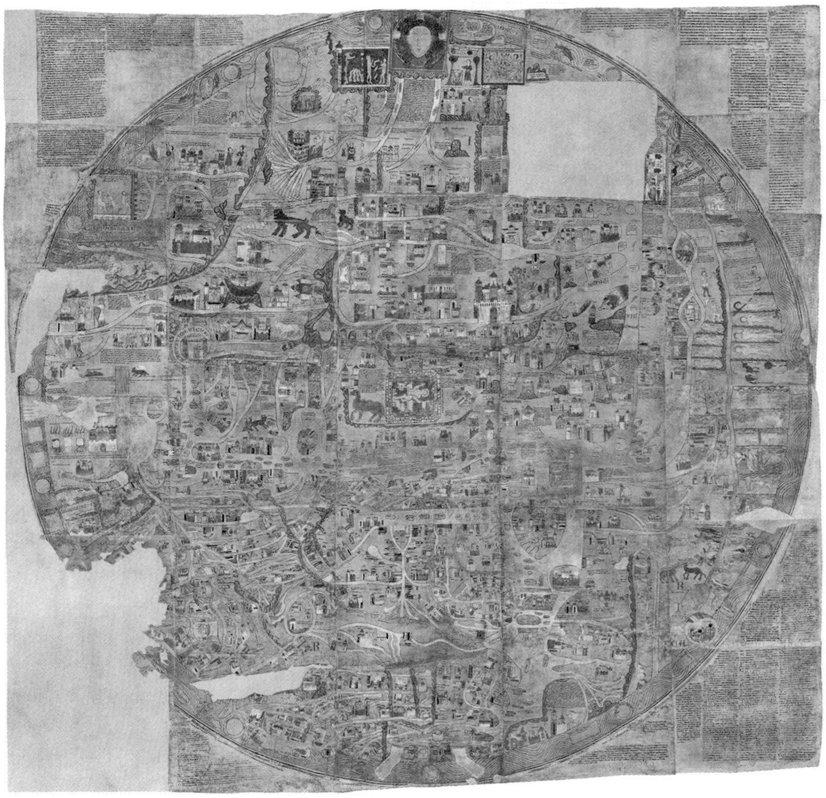

6. Ebstorf world map, late 13th century; east at the top. Reconstruction by Hartmut Kugler of the large encyclopaedic map destroyed in an air raid on Hanover in 1943; it was probably copied from a map of about 1239. Berlin, Akademie Verlag.

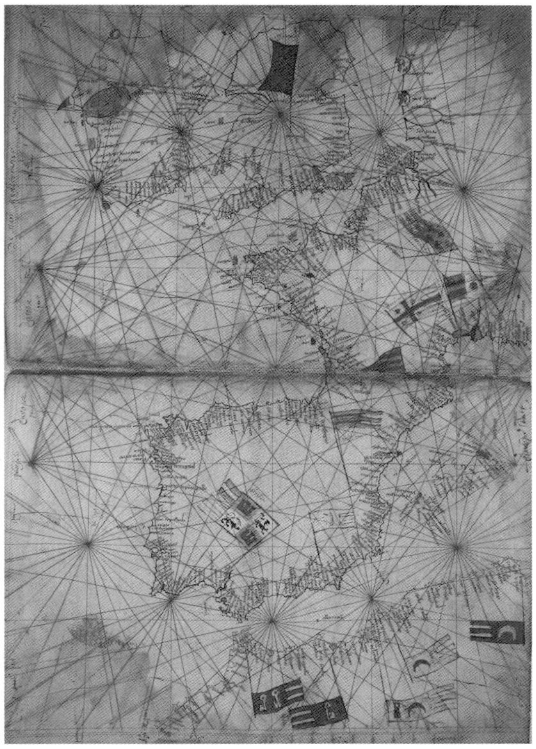

7. Western Europe on a portolan chart by Pietro Vesconte of Venice, 1330. While we recognise most of the outlines, the coasts beyond the Thames estuary and Flanders, less well known to Italian seamen, are entered only sketchily. BL, Add. MS 27376*, ff. 180v-181r.

the mid-fourteenth century – and they had far-reaching influence on late-medieval cartography in general.

Before the fifteenth century Palestine was the only area for which we have a significant number of regional maps; the few others, predictably, are from north-west Europe and Italy. The oldest, of about 1200, is a map of the British Isles by Gerald of Wales (Giraldus Cambrensis) of which we have four copies; its rudimentary outlines, perhaps also the few places named, probably came from a world map that showed the ocean islands in stylised form.[23] More sophisticated are the maps by Matthew Paris, monk of St Albans who died in 1259; besides the maps of Palestine that we shall look at in detail, he drew four maps of Britain. They are probably two separate compositions, one of them in three versions; each of the two took its outline from a different world map, with internal detail added from other sources, a technique that we shall meet several times in maps of Palestine.[24] A later, more enigmatic, map of Britain is the Gough map, drawn perhaps as late as the early fifteenth century, perhaps from a late-thirteenth-century exemplar. Here the outline of south-east England looks relatively familiar but the rest of the coast does not, just as on contemporary portolan charts, but there is no obvious model for the network of lines joining the many towns with distances between them.[25] Our only regional maps of Italy are both in a single manuscript, a copy of Paolino Veneto's *Great chronology* (*Chronologia magna*) produced at Naples, a manuscript that also included a map of Palestine.[26] One shows the whole of Italy on a single page, the other, larger, map divides the peninsula at Rome; the two halves are side by side on a single opening **[8]**. Both clearly come from portolan charts, with inland features added, though curiously each has south, not north, at the top.[27]

None of the regional maps of Palestine to be considered in detail was first drawn after about 1300, though in some cases we shall be looking at much later copies. This, then, was the cartographic context in which they were created.

23 Described in *Giraldi Cambrensis opera*, ed. by J.S. Brewer et al., Rolls Series 21 (8 vols, London, 1861-91), v, pp. xviii-xix; further references are given by Thomas O'Loughlin, 'An early thirteenth-century map in Dublin: a window into the world of Giraldus Cambrensis', *Imago Mundi*, 51 (1999), p. 34, n.16.

24 *Four maps of Great Britain*, ed. Gilson; Harvey, 'Paris's maps of Britain'.

25 Nick Millea, *The Gough map: the earliest road map of Great Britain* (Oxford, 2007); T.M. Smallwood, 'The date of the Gough map', *Imago Mundi*, 62 (2010), pp. 3-29.

26 Map I in chapter 10, pp. 120-1.

27 Discussed and illustrated by Degenhart and Schmitt, 'Sanudo und Veneto', pp. 84-5, 128-30.

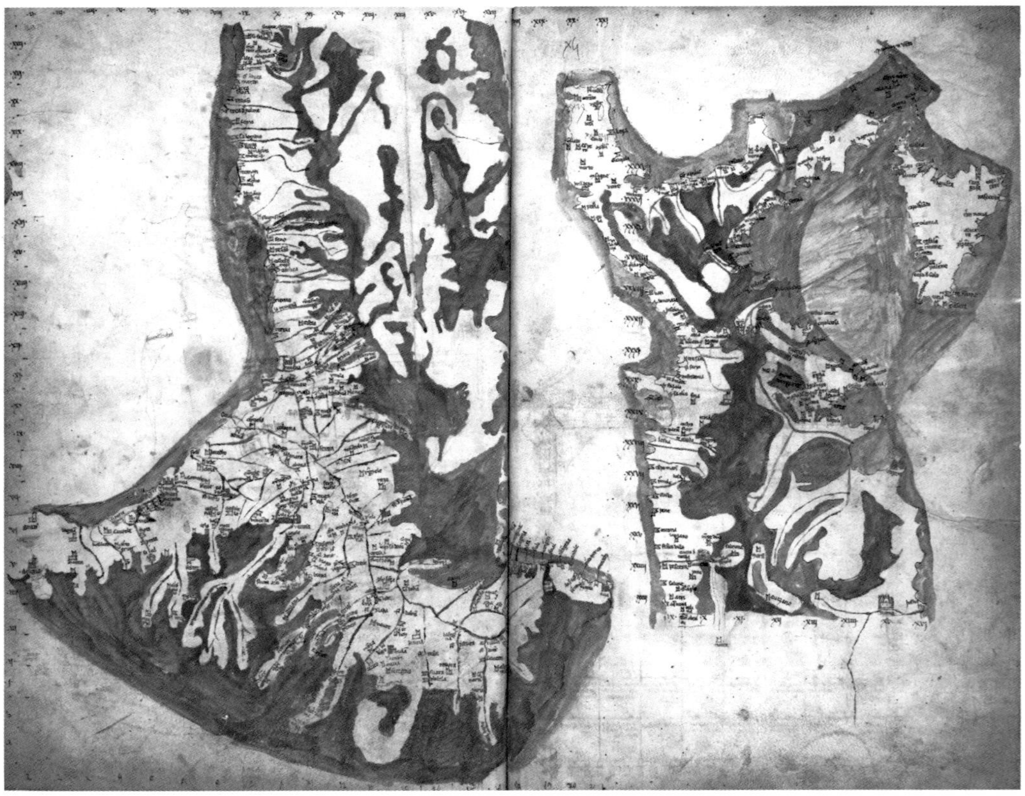

8. Map of Italy, drawn at Naples, 1334 × 1339; south at the top. The map is divided in two, with the northern half on the left; Sicily, on the right, is turned sideways to fit the page. Vatican City, Biblioteca Apostolica Vaticana, ms Vat. lat. 1960, ff. 267v-268r.

MAPS OF THE HOLY LAND

4

In the following chapters we shall look in detail at the eight surviving regional maps of Palestine that were first created between the late twelfth century and about 1300. Others may have failed to survive; thus, the map drawn on cloth that accompanied a copy of the late-thirteenth-century treatise of Galvano da Levanto may or may not have been a version of one of the maps we know.[1] In any case, our regional maps are not the only maps covering all or part of the area that survive from the Middle Ages, and in this chapter we shall look, much more briefly, at the others that show all or much of the region.

First are three maps, all older than those we shall examine in detail and so far as we can detect quite unconnected with them – or with each other. However, in the form that we have it, the earliest of the three, the Peutinger map or table, is contemporary with the Ashburnham Libri and Tournai maps, for it dates from about 1200, but it is a copy of a map that may have been first constructed as early as the second century and probably not later than the fourth. It is the nearest thing to a Roman world map that survives to us, showing the whole empire with its coasts, rivers, roads, towns and other staging posts with the distances between them – but all squeezed from top to bottom to form a strip probably originally some 8 metres long but only 33 centimetres high. This means that, while the map shows places along the network of routes in correct sequence, the shapes of lands and coastlines are unrecognisably distorted. It most likely comprised fourteen sheets of parchment glued together, and extended from Britain on the left to India on the right, with Rome in the centre, but the three westernmost sheets are lost. The map takes its name from its sixteenth-century owners, Konrad Peutinger of Augsburg and his heirs; a half-size engraving of the entire map was published in 1598 and since then it has been published and discussed in a number of successive studies. Richard Talbert published a full analysis and discussion in 2010 but, as he makes clear, some of his conclusions can only be tentative.[2]

The map's distortions mean that Palestine appears with west at the top, looking across a narrow gulf to Anatolia, which is oriented to the north **[9]**. In the area from the Mediterranean as far east as Damascus and from the Nile as far north as Antioch some eighty places are named, and twenty-three of them are marked with pictures of buildings that have been identified as conventional signs – in the whole map five different forms have been distinguished. All but one of those in Palestine have two pointed towers, by far the most common form throughout the map, meaning probably an administrative centre or simply a place where lodging is to be had; however at *Addianam*, near Mount Sinai, a substantial building without towers has been taken to mean a place where there was room for herds of animals.[3] At Antioch

1 It was described both as a map of the Promised Land and as a map of the kingdom of Jerusalem (Franciscus Ehrle, *Historia bibliothecae romanorum pontificum*, vol. 1 (Rome, 1890), pp. 90, 104; Gautier Dalché, 'Cartes, réflexion stratégique et projets', pp. 83-4).

2 Talbert, *Rome's world*. Earlier works include Levi, *Itineraria picta*; Luciano Bosio, *La Tabula Peutingeriana: una descrizione pittorica del mondo antico*, I Monumenti dell'Arte Classica 2 (Rimini, 1983); the account by O.A.W. Dilke in *History of cartography*, pp. 238-42; and the discussion of its origin by Benet Salway, 'The nature and genesis of the Peutinger map', *Imago Mundi*, 57 (2005), pp. 119-35.

3 Levi, *Itineraria picta*, pp. 66-85; Talbert, *Rome's world*, pp. 118-21.

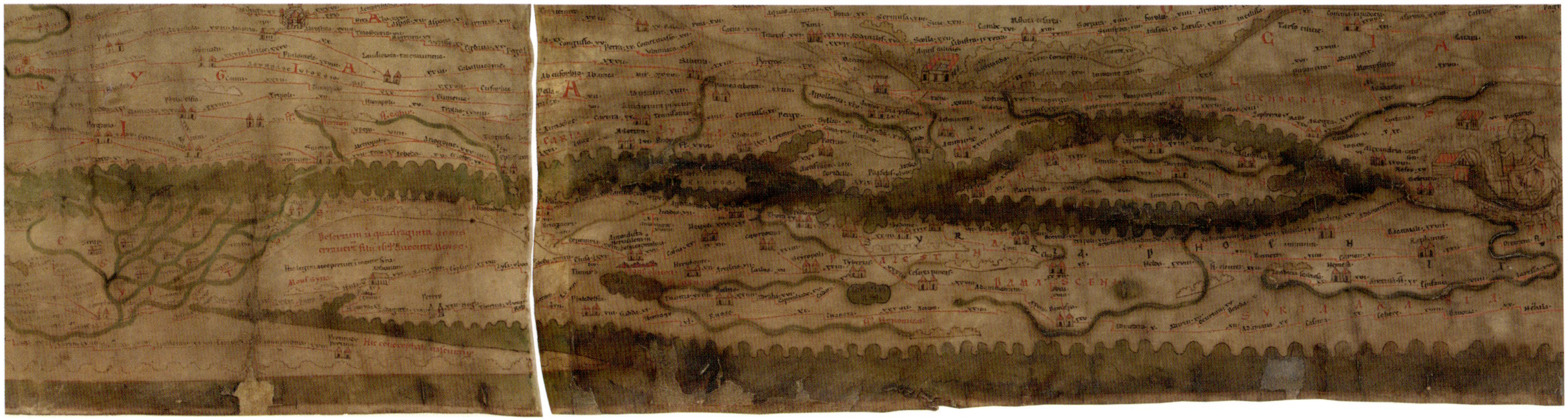

9. Palestine on the Peutinger map, a copy, about 1200, of a 2nd- to 4th-century original. Directions are badly distorted so as to show routes throughout the Roman empire on a long strip. In this section Palestine is at the foot, its coast looking across to Anatolia; on the left is the Nile delta, on the right Antioch shown as a seated figure. Vienna, Österreichische Nationalbibliothek, Codex Vindobonensis 324.

there is no conventional sign but a seated figure in Roman dress with helmet and spear, like the symbols that mark Constantinople and Rome. Talbert argues, convincingly, that this symbol, originally used at Rome alone, was added when the map was being copied at some point in the Christian era – Antioch and Constantinople were patriarchates – and in Palestine we see other added Christian elements: the mount of Olives is named, and Mount Sinai, and the desert where the children of Israel wandered for forty years.[4] However, there is no special symbol at Jerusalem or Alexandria, the other two patriarchates, perhaps simply for lack of space, and the places named in Palestine belong firmly to the pre-Christian period: Bethlehem, Nazareth, Jericho and Capernaum do not appear and Jerusalem is named as 'Formerly Jerusalem, now Aelia Capitolina' (*Antedicta Herusalem modo Helya capitolina*). For records related to the Peutinger map, however distantly, we must look to the itineraries of the Roman period, not to the maps of the Middle Ages.

The Peutinger map was well known long before scholars first heard of the Madaba mosaic map in 1897 **[10]**.[5] Set in the floor of St George's church at Madaba in Jordan, it is likely that it was originally nearly 7 metres across and over 22 metres from side to side. Its inscriptions are in Greek and it probably dates from the early seventh century, when Madaba was of greater importance than now; indeed, the town was long deserted and the mosaic came to light only after it had been repopulated in 1880. Much of the map is lost, but along with three tiny fragments there remains an irregularly shaped area 5 metres across and 10.5 metres wide, enough to give a clear idea of the map's character and scope. With east at the top, it extended certainly to the Nile in the south, well beyond the Jordan in the east and probably to Damascus and Byblos in the north; in the parts that survive we see much of the Nile delta, the Dead Sea, the southern end of the Jordan and Jerusalem with its surrounds to north and west, and we get tiny glimpses of the Mediterranean coast. It is elaborate and detailed, and it is reasonable to suppose that other mosaics to the same design were laid in other centres.

Basically it is a picture map: mountain areas are shown as outlines filled in with variegated colours, but all else is pictorial. There are waves on all waters, fish swim in the rivers and on the Dead Sea are two boats, with figures effaced deliberately by being replaced with random tesserae. There is a ferry across the Jordan[6] and beyond it a deer is being hunted. Villages are marked by symbolic buildings and an attempt has been made to classify them as conventional signs, as on the Peutinger map. Nine towns, however, appear as detailed agglomerations of houses, and their main streets, as at Gaza, Ascalon and above all Jerusalem, are opened out so that we see the arcading on each side. Some buildings and monuments are recognisably copied from life, certainly at Jerusalem and Nablus and probably in the other towns too. As we would expect from its location in a church, the mosaic is above all a map of places named in the Bible and many of the inscriptions spell out their biblical associations: 'Bethzachar, the place of Saint Zacharias', 'Modeim, today Moditha, the place of origin of the Maccabees' and many more. Each tribe is named, in large red letters, in its allotted area, a possible link with later world and regional maps.

It is certainly a link, though at best an indirect one, with our third early regional map of Palestine; this is entitled 'Map of the Land of Promise' (*Item figura terre*

4 Talbert, *Rome's world*, pp. 106, 124.

5 The earliest substantial work on the map, Adolf Schulten, *Die Mosaikkarte von Madaba und ihr Verhältnis zu den ältesten Karten und Beschreibungen des heiligen Landes*, Abhandlungen der Königlicher Gesellschaft der Wissenschaften zu Göttingen, Philologisch-Historische Klasse, new series 4 (Berlin, 1900), examines its possible relationship with the later medieval maps of Palestine. M. Avi-Yonah, *The Madaba mosaic map* (Jerusalem, 1954) reproduces a drawn coloured copy of the map with accompanying description and discussion; Vilnay, *Holy Land*, pp. 2-3, 85, 93, 95, 120, 200, illustrates and discusses the map's depiction of Jerusalem and other places; O.A.W. Dilke gives a brief account of the map in *History of cartography*, pp. 264-5; and *The Madaba map centenary 1897-1997: travelling through the Byzantine Umayyad period*, ed. by Michele Piccirillo and Eugenio Alliata (Jerusalem, 1998), is a collection of short studies on particular aspects of the map. The inscriptions are translated on a redrawn copy of the map in Wilkinson, *Pilgrims before the crusades*, pp. 152-6.

6 I owe its identification to Dr Catherine Delano Smith.

10. Madaba mosaic map of Palestine, probably early 7th century; east at the top. Only fragments survive of a map originally 22 metres wide. Above is part of the largest fragment, showing the Dead Sea and Jerusalem; both at Jerusalem and at Gaza, on the right, we see the arcading on each side of the main street. Madaba, St George's church.

repromissionis) and it shows the division among the twelve tribes in the book of Joshua **[11]**. This came to light even later than the Madaba map; it was first noticed in print in 1954 and first discussed in detail by Thomas O'Loughlin in 2005.[7] It is appended to the section on Joshua in a commentary on the Bible entitled *Some problems among the difficulties of the canonical texts* (*Pauca problesmata de enigmatis ex tomis canonicis*). This work was written before 800 in Ireland or in a religious house on the Continent with strong Irish connections, but the map is in a mid-ninth-century copy that was possibly written in France. Only one other complete copy of the text is known, and while this lacks the map it includes a relevant heading, suggesting that the copyist simply left out the map that was in his exemplar. The map we have is the work of a copyist, not the original mapmaker – thus *Ramoth* and *Galaad*, apparently shown as separate places, are really the single Ramoth-Gilead – and O'Loughlin argues that the map was an integral part of the work from the start. A map was a neat way of setting out the division of Canaan among the tribes, cutting through the various textual difficulties and inconsistencies of the book of Joshua.

The wording on the map is in what now appears as grey-brown, but a redder shade is used for the title and for the other features. North is at the top, so the Mediterranean coast, drawn as a thick line, is on the left. The Jordan flows from a triangular source through the sea of Galilee (unnamed) and, turning west, through the Dead Sea into a river that leads into the Mediterranean through the land of the Philistines. The areas of the tribes are bounded by straight lines. Eight cities are marked with what is probably the ground plan of a square enclosure with its entrance on the east side, while the symbol for Jerusalem, bigger than the rest, has pointed triangles along two sides, perhaps, O'Loughlin suggests, a hint of Mount Sion; further cities – Ramoth, *Galaad* and Jaffa – are named without a symbol, as well as the site of Jericho, destroyed by the Israelites and accursed by Joshua. The cities include the six named

7 Bernhard Bischoff, 'Wendepunkte in der Geschichte der lateinischen Exegese im Frühmittelalter', *Sacris Erudiri*, 6 (1954), p. 226, where it is a little unkindly called 'an attempt at a map' ('ein Versuch einer Karte'); Bischoff's work was translated as 'Turning-points in the history of Latin exegesis in the early Middle Ages', in *Biblical studies: the medieval Irish contribution*, ed. by Martin McNamara, Proceedings of the Irish Biblical Association 1 (Dublin, 1976), where the reference to the map is on p. 99; Thomas O'Loughlin, 'Map and text; a mid ninth-century map for the book of Joshua', *Imago Mundi*, 57 (2005), pp. 1-22, plate 1. The map is also reproduced, and briefly described, in *Map book*, pp. 40-1.

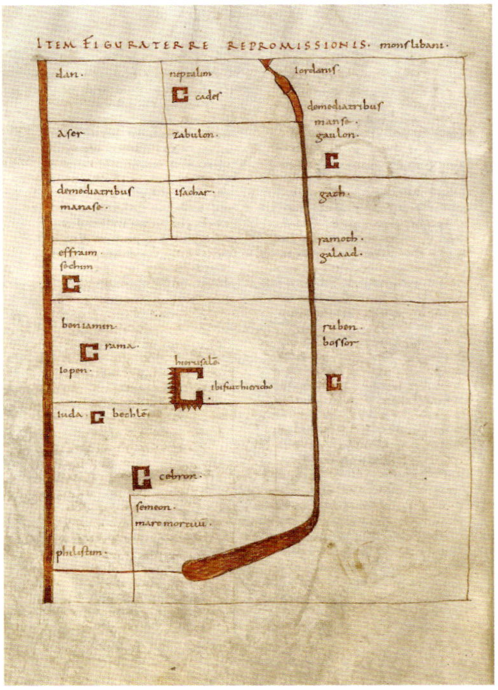

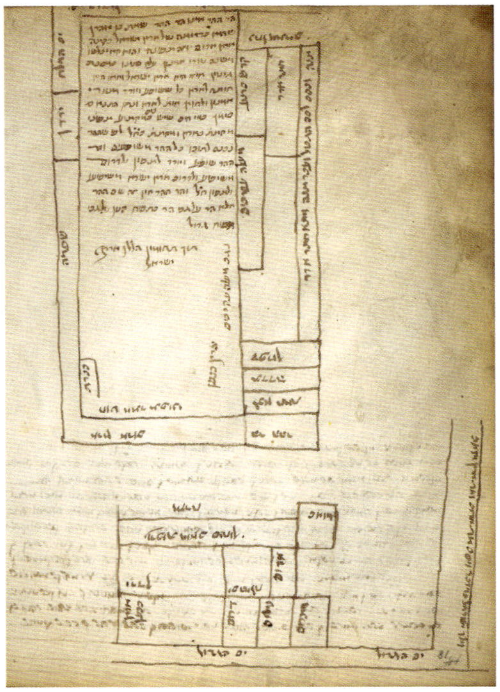

11. Map of the Promised Land, mid-9th century; north at the top. Accompanying a commentary on the book of Joshua, the map shows the country divided between the twelve tribes. At the foot a stream links the Dead Sea with the Mediterranean. Paris, Bibliothèque nationale de France, ms latin 11561, f. 43v.

12. Maps of Canaan in the tradition of Rashi, 14th century. The upper map, with south at the top, shows features around the borders of Canaan; the lower, with east at the top, includes Moab, Egypt and further areas beyond the borders. BL, Harley MS 5529, f. 18r.

in the book of Joshua as places of sanctuary for anybody who killed someone without intent.[8] This, O'Loughlin shows, points to the mapmaker's Irish interests, for there is a relevant chapter in an early-eighth-century collection of ecclesiastical rules that was written in Ireland.

This map of the Land of Promise is the oldest known map of Palestine to illustrate a commentary on any part of the Bible, but others followed. Maps and other diagrams accompany the Hebrew commentary on the Jewish Bible and the Babylonian Talmud (which set out Jewish laws and ceremonies) by Rabbi Shlomo Itzhaki; known as Rashi from the initial letters of this name, he was born at Troyes in 1040 and died there in 1105. The illustrations were first drawn by Rashi himself, but they are not all to be found in all of the many surviving manuscripts of his work.[9] Three are maps of Palestine. One sets out the boundaries of Canaan and another the lands beyond its borders, both illustrating its description in the book of Numbers **[12]**.[10] The third shows the idealised division between the tribes envisaged by Ezekiel on their return from the Babylonian exile.[11] All three are made up of stylised, mostly rectangular, shapes, simply showing the relationship of one place to another, though some manuscripts present them in more decorative form. They may well owe something to early-medieval Christian maps.[12]

Catherine Delano Smith has shown how two later Christian commentators drew on Rashi's maps and other diagrams; both, like Rashi, lived and worked in northern France.[13] The first was Richard of Saint-Victor, author of a commentary on the book of Ezekiel, who died in 1173; illustrations adapted from Rashi's, among them the map of Ezekiel's idealised Canaan, are to be found in all twenty-two manuscripts known of his work except one, where spaces for them have been left blank. Copies of Richard's map of Ezekiel's division of Canaan, as well as a map of the tribal division in the book of Joshua,[14] have been added by a later hand to a mid-thirteenth-century Bible that was written in England and is now at Brasenose College, Oxford; the maps appear first in the margins of the relevant texts, then again, drawn larger, on blank leaves at the end **[13]**.[15] The map of the book of Joshua's division of Canaan appears again in a thirteenth-century copy on a leaf attached to a contemporary Bible in Trinity College, Cambridge.[16] The other, much later, commentator, was Nicholas of Lyra, who died in 1349. He wrote two commentaries on the whole Bible, one on its literal the other on its mystical meaning (*Postilla litteralis*, *Postilla mystica*), and it is in manuscripts of the former that we find illustrations derived from Rashi, among them again the map of Ezekiel's idealised Canaan – unsurprisingly, for Nicholas cites and discusses Rashi's text. Many manuscripts survive – Nicholas's work was often copied – but not all include all, or indeed any, of the illustrations.[17]

None of these biblical commentaries gives us a plan of Jerusalem, but in the twelfth century plans of the city appear in other contexts, among them chronicles of the crusades and the encyclopaedic *Floral book* (*Liber floridus*) by Lambert of Saint-Omer.[18] One, known from a single manuscript, shows the city as an irregular quadrilateral, reflecting direct knowledge **[14]**, and it can be seen as the precursor of later realistic plans of Jerusalem and other cities in Palestine:[19] Acre on Matthew Paris's mid-thirteenth-century map (chapter 8, pp. 82-3) and the plans of Jerusalem, Acre and Antioch that accompany some of the fourteenth-century grid maps

8 Joshua 20.7-8.

9 Mayer I. Gruber, 'What happened to Rashi's pictures?', *Bodleian Library Record*, 14 (1991-4), pp. 116, 121.

10 Numbers 34.2-12.

11 Ezekiel 48. A further plan shows the layout of the sanctuary, as described in Ezekiel 45.

12 Copies in manuscripts from the thirteenth to the fifteenth century are reproduced and described by E. and G. Wajntraub, *Hebrew maps of the Holy Land* (Vienna, 1992), pp. 2-18, and by Catherine Delano Smith and Mayer I. Gruber, 'Rashi's legacy: maps of the Holy Land', *The Map Collector*, 59 (1992), pp. 30-5, as well as in what is the fullest account of these maps in English, Benjamin Z. Kedar, 'Rashi's map of the land of Canaan, ca.1100, and its cartographic background', in *Cartography in Antiquity and the Middle Ages*, ed. by Richard J.A. Talbert and Richard W. Unger, Technology and Change in History 10 (Leiden and Boston, 2008), pp. 155-68.

13 Catherine Delano Smith, 'The exegetical Jerusalem: maps and plans for Ezekiel chapters 40-48', in *Imagining Jerusalem*, ed. Donkin and Vorholt. I am most grateful to Dr Delano Smith for crucial guidance on Rashi's maps and their later derivatives.

14 Joshua 13-19.

15 Oxford, Brasenose College, MS 5, ff. 32r, 281r, 442r,v; Coxe, *Catalogus codicum*, II, Brasenose College, pp. 1-2; Alexander and Temple, *Illuminated manuscripts*, p. 20. I am grateful to Dr Lesley Smith for advice on these maps.

16 MS O.9.13, f. iiir. This is the map mentioned by Montague Rhodes James, *The western manuscripts in the library of Trinity College, Cambridge: a descriptive catalogue* (3 vols, Cambridge, 1900-2), III, p. 453, as 'a rude map of the Holy Land shewing the positions of the tribes'.

17 Nicholas of Lyra's illustrations are discussed in general by Lesley Smith, 'The imaginary Jerusalem of Nicholas of Lyra', in *Imagining Jerusalem*, ed. Donkin and Vorholt.

18 All except the ink sketch discussed by Hanna Vorholt (below) and BL Add. MS 32343, f. 15v (Harvey, *Medieval maps*, p. 90), are reproduced and discussed by Simek, 'Hierusalem', pp. 121-53.

19 It is discussed in detail by Vilnay, *Holy Land*, pp. 8-9, and by Heydenreich, 'Jerusalem-Plan'.

(chapter 10, pp. 109-10, 113-14). Other plans of Jerusalem that originated in the twelfth century stem from three distinct archetypes, but they are all rather further from reality, showing the city in more or less stylised form within a circular wall.

Most of these city plans show something of the immediate surrounds, but copies of two of the three stylised plans of Jerusalem show a good deal more. Of one we have only two copies, recently shown by Hanna Vorholt to be the same map; they look quite different, for one is simply sketched in ink while the other is carefully drawn and coloured in stylised form with names placed in small circles **[15]**.[20] Within Jerusalem nothing is marked but Calvary and the Holy Sepulchre, the Temples of Solomon and of the Lord,[21] and the Hospital of St John. From each of its four named gates lines of routes connect the city with Acre, Caesarea Palestina, Jaffa and Ascalon on the coast, with Bethlehem to the south and with Jericho, the Jordan and the Dead Sea, as well as with more than twenty other places, mostly sites of biblical events and of pilgrimage: 'Bethany where Lazarus was brought to life' (*Bethania ubi Lazarus est suscitatus*), 'pool of Siloam' (*Aqua siloe*) and so on. The other plan is known from eleven copies; they are all broadly alike in the street plan and buildings of the city, but while three mark only a few nearby sites outside the city wall, the rest extend a good deal further. None shows the coast, but the Jordan appears on three of them, of which one, curiously, shows the sea of Galilee, the sea of Tiberias and Lake Gennesaret as three separate bodies of water **[16]**.[22] All are pictorial: buildings, towns and mountains are drawn, in more or less stylised form, as seen from the side.

20 Hanna Vorholt, 'Studying with maps: Jerusalem and the Holy Land in two thirteenth-century manuscripts', in *Imagining Jerusalem*, ed. Donkin and Vorholt.

21 I.e. the El Aqsa mosque and the Dome of the Rock.

22 Simek, 'Hierusalem', p. 134.

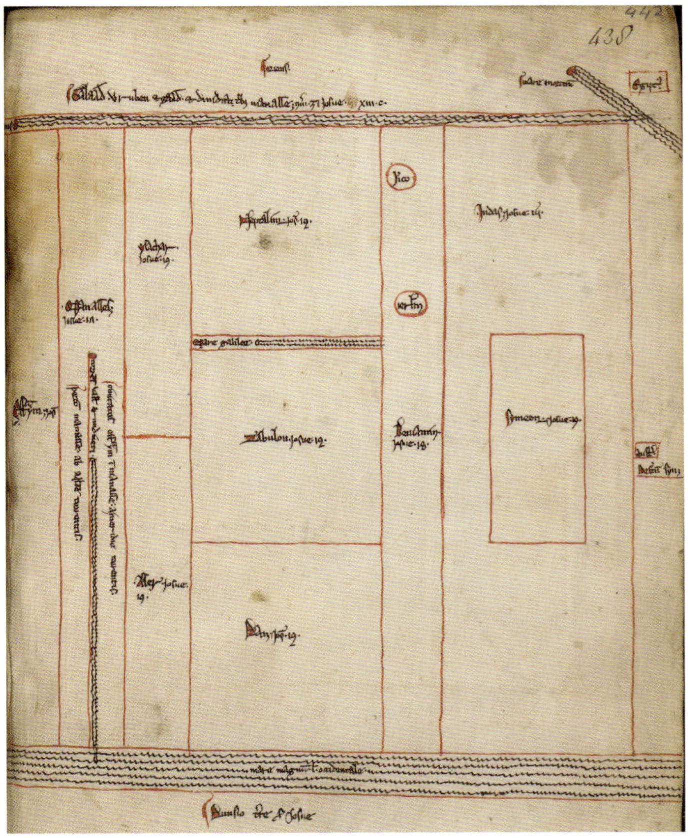

13. Map of Canaan, late 13th century; east at the top. The lines mark the tribal division of the book of Joshua. The Dead Sea, top right, is diagonal to the River Jordan. Oxford, Brasenose College, MS 5, f. 442r.

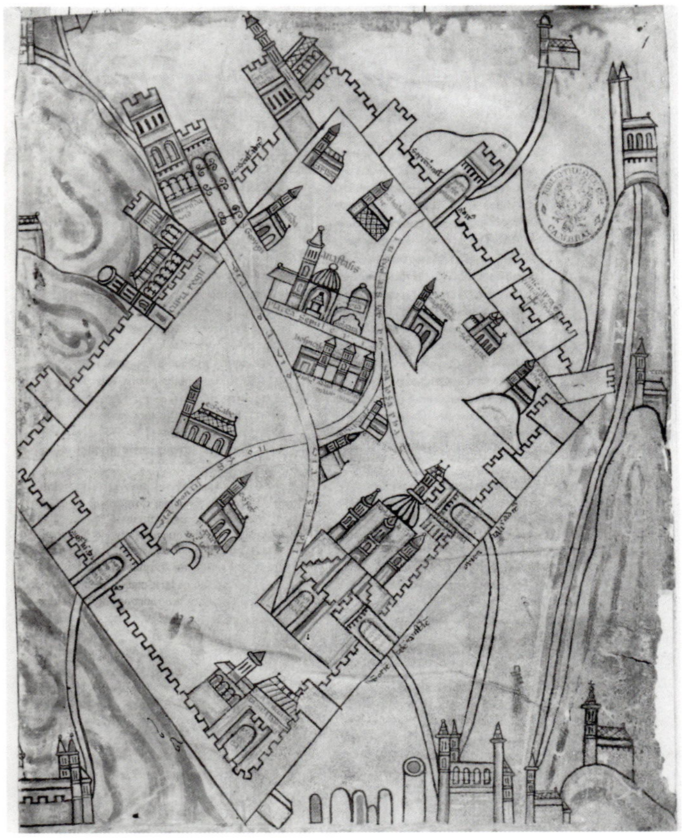

14. Plan of Jerusalem, mid-12th century; north-west at the top. The earliest plan to show the city as a quadrilateral, it identifies the tower of Tancred d'Hauteville, the house of the Templars and other contemporary features. Cambrai, Centre culturel, ms 437, f. 1r.

Only a few of these early plans of Jerusalem extend far beyond the city, but the encyclopaedic world maps of the thirteenth century give us full and detailed maps of all Palestine. However, their two early precursors, the eleventh- and twelfth-century Cotton and Munich maps, tell us little, though the area is shown disproportionately large near the centre of each.[23] There are some similarities between them. Both show Jerusalem near the coast, Bethlehem, Jericho, Caesarea Philippi and Antioch and, though Tyre is not named on the Cotton map, both show the peninsula on which it stood; the Cotton map also names Hebron, the Munich map Nazareth and Capernaum, both placed east of the Jordan. The physical geography is closer to reality on the Cotton map, which gives Palestine a fairly straight coastline and shows the Jordan ending in what must be the Dead Sea, whereas on the Munich map the Jordan, with no lakes, flows into the Red Sea and for the first time, but by no means the last, we see a great gulf that gives Palestine a northern as well as a western coast. On the Cotton map straight lines that bound provinces in Europe, western Asia and Africa here mark the areas of the twelve tribes – nine of them are named. The twelve tribes with straight bounds reappear on the Sawley map of about 1200, the only one of our world maps that names all twelve.[24] Here the coastline has a gulf to the south of Palestine as well as to the north, while the Jordan flows through the sea of Galilee before ending in the Dead Sea; two tributaries are shown, the Kishon (unnamed) and the Arnon as well as the Wadi El Arish (*Rinocorua*) bordering on Egypt. The mountains of Lebanon and of Gilead are shown as great ranges, unlike the isolated peaks on the earlier maps, Jerusalem appears inland, and the other towns marked are Bethlehem, Jericho, Beersheba, Ascalon, Caesarea Philippi, Tyre (without its peninsula), Sidon, Damascus and Antioch.

When we move from their precursors to the three great encyclopaedic world maps

23 There are good reproductions of both the Munich and the Cotton maps in Harvey, *Medieval maps*, pp. 22, 26, and of the Cotton map in *Map book*, p. 47.

24 The best reproduction is in *Map book*, p. 51.

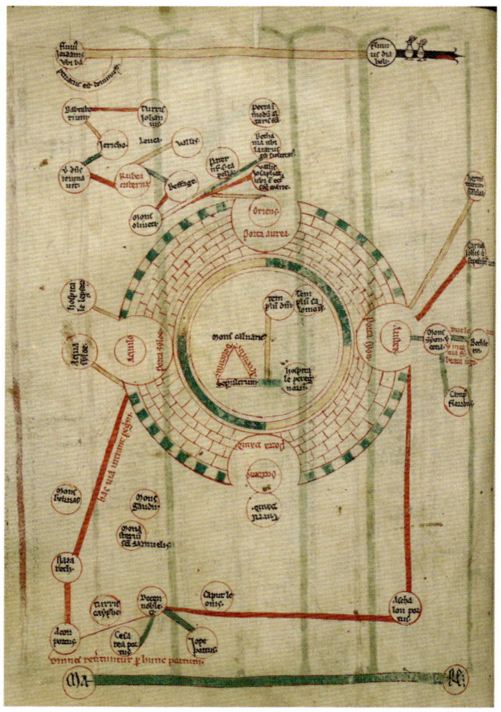

15. Plan of Jerusalem and surrounds, first half of the 13th century; east at the top. The plan extends from the sea to the Jordan with, top right, the Dead Sea, 'fluuius diaboli' ('the Devil's river'), shown with two jars giving off malodorous fumes. Lines show routes between the places named in circles. BL, Harley MS 658, f. 39v.

16. Plan of Jerusalem and surrounds, second half of the 12th century; east at the top. Eleven copies are known of this stylised circular plan of the city, varying a good deal in what they show outside the walls. Here the sea of Galilee appears at the top as three lakes, each with a different name. Brussels, Bibliothèque royale de Belgique, ms 9823-9824, f. 157r.

that we know in detail, we meet vastly fuller maps of the area.[25] If we extend as far as the Nile our tally of towns and other local features named on each map, we find no more than 14 on the Sawley map and rather fewer on the Cotton and Munich maps, whereas on the Vercelli map there are 72, on the Ebstorf map 93 and on the Hereford map 58.[26] There are in fact more places named in Palestine on the Ebstorf map than on any of the regional maps we shall examine that are earlier than the large Burchard map at the end of the thirteenth century. In their general appearance and in their detail the three maps are very different from each other. At the same time, as we shall see, they have some features in common **[17, 18, 19]**. On each the coastline turns east at some point north of Acre, giving Palestine a northern coast; on Hereford, as on the Sawley map, there is a similar gulf to the south, and on Ebstorf a channel from the Red Sea to the Mediterranean completely separates Palestine from Egypt. On all three the Jordan rises on Mount Lebanon (*mons libanus*, *mons lybanus*), on Vercelli, unusually, from a single source but on the other two in separate sources for the Jor and the Dan. Ebstorf omits the sea of Galilee unless as a slight widening of the river, but all three show the Dead Sea, Vercelli and Ebstorf with descriptions, textually unrelated to each other, Ebstorf and Hereford with pictures of the cities of the plain submerged below the water. On Hereford the Jordan has no outlet, on Vercelli it joins a river that runs into the Mediterranean, while on Ebstorf the Dead Sea immediately adjoins the Red Sea and may be supposed to flow into it. On all three maps mountains appear either as single humps or else as small clusters (Vercelli) or long ranges (Ebstorf and Hereford).

All three maps mark towns with pictures of buildings, on Vercelli a uniform low structure with a squat tower, on Ebstorf and Hereford by a great variety of forms,

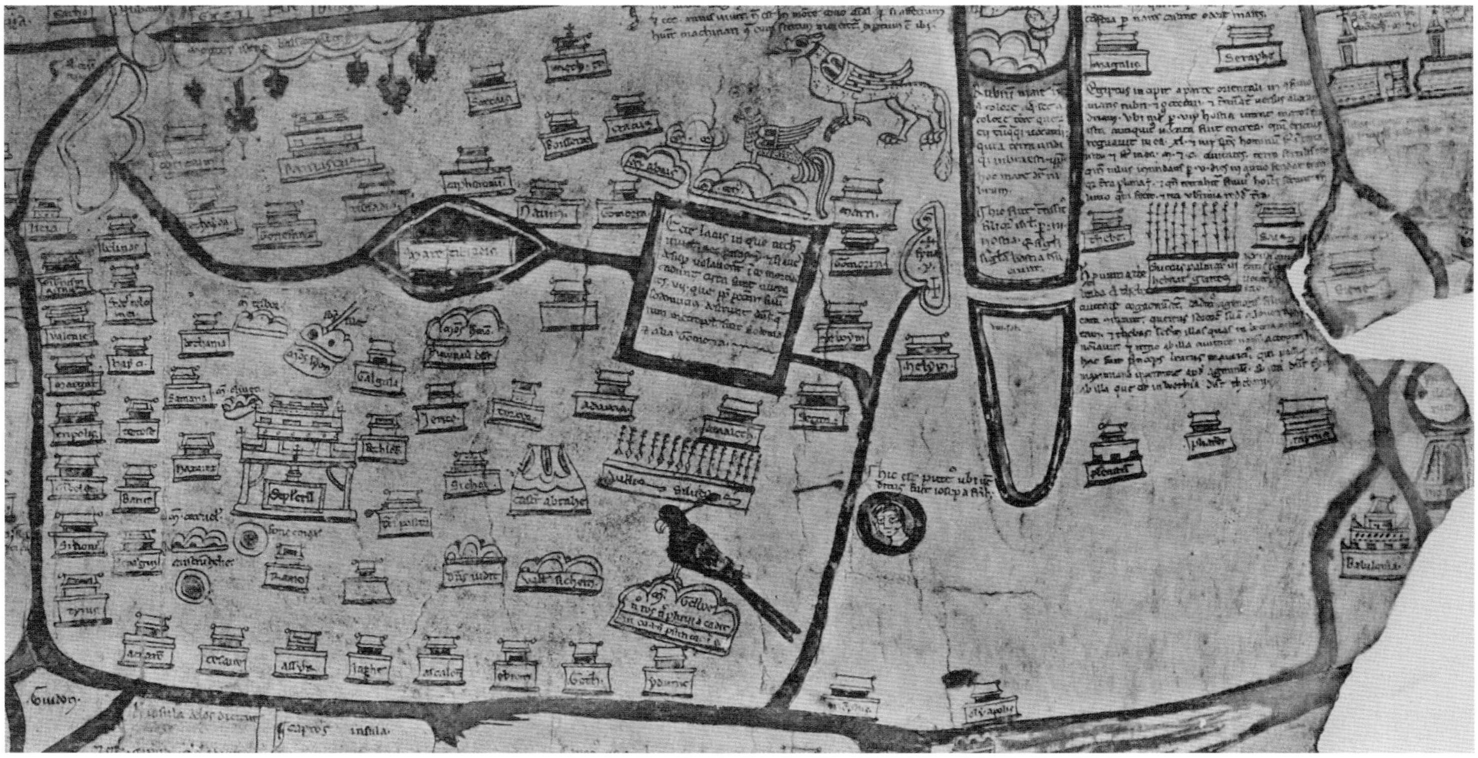

17. Palestine on the Vercelli world map, early 13th century; east at the top. From the lozenge-shaped sea of Galilee and the square Dead Sea the Jordan flows into the Mediterranean. At Jerusalem, otherwise unnamed, is a tomb, 'Sepulchre' ('Sepulcrum'), below a roof supported by pillars. Vercelli, Biblioteca capitolare.

25 The fullest accounts and reproductions are in Capello, *Mappamondo di Vercelli*; Kugler, *Ebstorfer Weltkarte*; and Westrem, *Hereford map*.

26 P.D.A. Harvey, 'The Holy Land on medieval world maps', in *Hereford world map*, ed. Harvey, p. 248.

18. Palestine on the Ebstorf world map, late 13th century; east at the top. Reconstruction by Hartmut Kugler. The Dead Sea, top right, with its brown waves and submerged cities, flows into the Red Sea. At Jerusalem on the left, Christ rises from the tomb, to be viewed from the left side. Berlin, Akademie Verlag.

19. Palestine on the Hereford world map, about 1300; east at the top. Jerusalem appears as a battlemented circle, with above it the figure of Christ on the Cross. In the Dead Sea, at the top, are the submerged Sodom and Gomorrah; beside and around it a twisting line shows the Israelites' wanderings in the desert. Hereford, Mappa Mundi Trust.

some, especially on Hereford, approaching the fantastic. Along the coast they are named in the realistic sequence that we shall see on the regional maps, though on both Vercelli and Hereford Acre appears twice, once as *Acharon* or *Acaron*, then again under its older name, *Sedes tolomei, Tholomaida*. Inland, however, the whole pattern is skewed by the displacement of Jerusalem, entered to the west of the sea of Galilee or even further north. This is common to all three maps, but is most obvious on Ebstorf. On Ebstorf and Hereford this was clearly to bring Jerusalem to the map's centre, the centre of the world's circle, another instance of the way geography was subordinated to symbolism on medieval world maps. On Vercelli, though Jerusalem lies on the map's central east-west axis, it is well to the east of the map's centre; the map may well have been based on one with Jerusalem at the centre, but the significance of this could have been overlooked – or disregarded – so as to give more space on the western half of the map, less on the eastern. That Jerusalem was moved to bring it to the centre of these maps, and not taken as the central point at the start of their construction, suggests that the idea of placing it there came late in the development of these world maps; on the Munich map Tyre is at the centre, on the Sawley map Caesarea Palestina.[27] The effect of moving Jerusalem varied on the three maps. On Vercelli and Hereford it drew other places northwards with it, so that, for instance, Bethlehem is still just a little to the south; on Ebstorf, however, Jerusalem with its immediate surrounds (Mount Sion, the pool of Siloam, the valley of Jehoshaphat, Bethany) moved north without affecting any other places, so that Bethlehem stayed behind at the other end of the country.

This portion of the Vercelli map offers little general information. Besides the description of the Dead Sea there is a short note on Mount Gilboa, where neither rain nor dew falls, with the picture of a bird that we shall return to in looking at Matthew Paris's Oxford map (chapter 7, pp. 70-1). There is an unnamed picture of a griffin, with the wings and talons of an eagle, the hindquarters of a lion, and Jerusalem is shown as an enlarged town symbol showing the Holy Sepulchre (*Sepulcrum*; the city is otherwise unnamed). On Ebstorf Jerusalem is shown with the square walls and twelve gates of the heavenly city and with a spirited picture of Christ stepping from the tomb, drawn to be viewed from the north;[28] beside the city is a camel, named *camelus*. Otherwise we are given only notes on particular places and regions: on Jerusalem, the Dead Sea, the two Caesareas and others.

On Hereford, however, there is rather more. Jerusalem here is circular and above it is the figure of Christ on the Cross; the distance from Mediterranean to Jordan has clearly been increased to accommodate this. In the south-east we see the phoenix, with a short description, and a yale, with a longer one – it was an animal with flexible horns – while near Gaza a mysterious *cirenus* bird is named but not described (the name may have been miscopied). No less mysterious is the otherwise unknown *marsok*, near Damascus, a beast whose four feet may each represent one of the four evangelists. Beside the Dead Sea is Lot's wife, turned into a pillar of salt, 'very forlorn-looking' according to W.L. Bevan and H.W. Phillott, authors of the first book on the map in 1873.[29] Even more dramatically the map shows the whole of the Israelites' Exodus from Egypt, drawn as a line starting near 'Joseph's barns' (*Orrea ioseph*, that is, the pyramids), crossing the Red Sea, wandering in the

27 Anna-Dorothee von den Brincken, *Fines terrae: die Enden der Erde und der vierte Kontinent auf mittelalterlichen Weltkarten* (Hanover, 1992), pp. 65, 183, sees a significance in Tyre's prominence on the Munich and other maps that is questioned by Patrick Gautier Dalché, '"Réalité" et "symbole" dans la géographie de Hugues de Saint-Victor', in *Ugo di San Vittore: atti del XLVII Convegno storico internazionale* (Spoleto, 2011), p. 373n.

28 Kugler, *Ebstorfer Weltkarte*, II, p. 171, notes possible reasons for this orientation.

29 Bevan and Phillott, *Mediaeval geography*, p. 77.

20. Palestine coast on a map of the Near East by Pietro Vesconte of Venice, 1330; east at the top. The coastline and the places named along it are those of the mapmaker's portolan charts. BL, Add. MS 27376*, ff. 182v–183r.

21. Palestine coast on a map of the Near East, drawn at Naples, 1334 × 1339. The places along the coast are from world or regional maps, with figures giving the distance from one to the next. Vatican City, Biblioteca Apostolica Vaticana, ms Vat. lat. 1960, f. 265v.

desert in a series of loops and finally crossing the Arnon and the Jordan to reach Jericho.

Our earliest portolan charts, contemporary with the Hereford map, are austere by comparison – and vastly more realistic. Along a broadly accurate coastline they enter a sequence of places that include those on the world maps together with others relevant to navigators. Thus on one anonymous chart of the early fourteenth century the names along the coast include Beirut, Jaffa and Ascalon but also *Capo Gloriata*, *Poio Conistabili* and others known to seamen but not to pilgrims.[30] This chart is unusual in sketching in the courses of some rivers; the earliest charts generally show no inland features.

The first chartmaker we know by name is Pietro Vesconte, who worked from at

30 Pujades i Bataller, *Les cartes portolanes*, p. 66–7.

latest 1311 in Genoa and Venice. We shall see that he was responsible for sets of maps that accompanied works by Marino Sanudo, among them the grid map of Palestine (chapter 10, pp. 112–13); among them too were charts of parts of the Mediterranean and Black Sea. One of the eastern end of the Mediterranean extended inland to include much of the Near East, as far as Cairo, the Persian Gulf and the Tigris **[20]**; in Palestine it shows Jerusalem, Nazareth, the Jordan and a very few other features and the line of names along the coast is that of the charts. We have nine copies of this map, drawn in Vesconte's workshop, dating from perhaps as early as 1316 until 1331 or 1332. As we shall see, we have also two copies of the map of Palestine that were produced in Naples between 1329 and 1339 to accompany the work of Paolino Veneto, and with them two copies of the map of the Near East **[21]**. These differ from the earlier version, however, in the details given inland, and the names along the coast are those not of the charts but of the world and regional maps, with the interesting addition of figures showing the distance between one place and the next.[31]

In turning his chart of the eastern Mediterranean into a map of the Near East, Vesconte was anticipating the way some charts of the whole Mediterranean came to include inland features, a style that seems to have developed in Majorca and that first appears in the charts of Angelino Dulcert in the 1330s. Features in Palestine that we see on one or other of these charts include Jerusalem with the Holy Sepulchre, Damascus, the Jordan rising in the mountains of Lebanon and flowing through the sea of Galilee to the Dead Sea, the place of Jesus's baptism and Mount Sinai with St Catherine's monastery – and some extend to the Tigris and beyond.[32] When a definitive study of the medieval maps of Palestine comes to be written it will need to take account of many maps besides the eight that we are about to examine in detail.

31 Degenhart and Schmitt, 'Sanudo und Veneto', pp. 22–6, 105, 115.

32 Pujades i Bataller, *Les cartes portolanes*, pp. 114–15, 120–1, 126–7, 152–3 (charts C7–C9, C13–C15, on the accompanying compact disk).

THE ASHBURNHAM LIBRI MAP

late 12th century[1]

5

Florence, Biblioteca Medicea Laurenziana, Ashburnham Libri ms 1882 **[22]**. Parchment, approximately 26.5 × 73.5 cm. Dark brown-black and red ink, with bright and dull blue, red, dark grey, mauve and green paint and gold leaf. Dorse not accessible.

IN 1846 Sir Frederic Madden, Keeper of Manuscripts at the British Museum, went to Paris to inspect the many manuscripts belonging to Guglielmo Libri, with a view to recommending their purchase by the Museum. Madden kept a voluminous diary, and he describes how he and his colleague John Holmes spent some two weeks examining all the manuscripts listed in a catalogue that Libri had sent them. Libri then produced a further sixteen items not in the catalogue – 'why omitted I do not know', wrote Madden. Madden numbered them and noted in his diary four of particular interest. Among them was '1882 A colored Map of Palestine on vellum, executed in the 12th century, about 2 ft by 10 in'.[2] This seems to be the earliest reference to the map that still bears the number that Madden gave it.

The Museum's projected purchase of the collection fell through. While Madden was discussing it with the secretary of the Museum a prominent London bookseller, Thomas Rodd, called on them. 'His account of Libri rather surprised and staggered me', Madden wrote. Rodd revealed that Libri was not simply a collector of manuscripts, as he had assured Madden, but was well known in Paris as a dealer. 'His character for probity was very problematical, and Rodd did not hesitate to say, that he had heard it repeated more than once, that Libri was suspected to have *stolen* a portion of the MSS. he possessed.'[3] This was in fact no more than the truth; moreover, Libri not only stole manuscripts but had them fraudulently altered. Some account of Libri, and an assessment of what happened to the map while it was in his hands, are essential preliminaries to discussion of the Ashburnham Libri map.

Guglielmo Libri, conte Carucci della Somaia, had a strange career.[4] Born in Florence in 1802, he was in the first instance a precocious mathematician and scholarly historian of mathematics. He was appointed to a chair of physics at the University of Pisa in 1823 but retired from teaching a year later, through bad health, while retaining his emeritus title and annual pension throughout his life. However, in Paris in 1840 he became professor of statistics (*calcul des probabilités*) at the Sorbonne – a post created especially for him – and between 1838 and 1841 he published the four volumes of his work, written in French, on the history of mathematics in Italy from the Renaissance to the end of the seventeenth century. He had had to leave Tuscany in 1831 because of his revolutionary sympathies; he settled in Paris and took French naturalisation. A keen and knowledgeable collector of books and manuscripts, he

1 I am grateful to Prof. P. Gautier Dalché for helpful discussion of this map and especially for telling me of its publication in 1992; to Prof. Richard Gameson and Dr Laura Nuvoloni for valuable advice on its handwriting; and to Prof. Sir Arnold Wolfendale for advice on other particular points.

2 Oxford, Bodleian Library, MS Eng.hist. c159 (a copy is BL MS Facs *1012/21), entry for 11 May 1846.

3 Ibid., entry for 3 June 1846.

4 Accounts of his life are in *Nouv. biogr. gén.*, XXXI, cols 129-31 (by 'J. et B.', unidentified); *Enciclopedia italiana* (36 vols, Rome, 1929-39), XXI (1934), pp. 67-8 (by Ettore Bortolotti); Giuseppe Fumagalli, *Guglielmo Libri*, ed. by Berta Maracchi Biagiarelli (Florence, 1963); and P. Alessandra Maccioni Ruju and Marco Mostert, *The life and times of Guglielmo Libri (1802-69): scientist, patriot, scholar, journalist and thief* (Hilversum, 1995).

22. The Ashburnham Libri map, late 12th century; east at the top. The map in its present state, as altered for Guglielmo Libri in the mid-19th century. Florence, Biblioteca Medicea Laurenziana, Ashburnham Libri ms 1882.

bought widely in France, Italy and England. After attempts to get a post in the Bibliothèque royale had failed, in 1841 he was appointed secretary to a newly set up commission to catalogue the manuscripts in provincial libraries throughout France. By the time he left the post in 1846 there were rumours of his thefts and when, in February 1848, he found that a report on his activities had reached the government he hurriedly moved to England. This report led to his trial *in absentia* in 1850 for stealing books and manuscripts; he was found guilty and sentenced to ten years' imprisonment. This prevented his return to France, and he continued in England, taking out naturalisation as a British subject. He claimed, and it was widely believed, that the charges against him were a political machination; he had been a friend of François Guizot, minister of King Louis-Philippe, whose fall from power, with the creation of the Second Republic, exactly coincided with Libri's flight. In 1869 he returned to Italy, but died in the same year at Fiesole.

The nearly two thousand manuscripts that Libri failed to sell to the British Museum in 1846 were bought in 1847 by the fourth Earl of Ashburnham, curiously through the agency of Thomas Rodd and John Holmes. By the time of the earl's death in 1878 the researches of Léopold Delisle, director of what was now the Bibliothèque nationale, had laid bare the scale and true nature of Libri's activities. He stole autograph letters from the Bibliothèque royale and other institutions in Paris, and he

took both complete manuscripts and single quires from the provincial libraries he was employed to inspect. This was easily done because of the deficient cataloguing: exactly what Libri's commission had been set up to rectify. He had letters he stole mounted in English style to give the impression they had been bought in England, and had manuscripts bound in wooden boards that pointed to an Italian origin. To the manuscripts he had *ex libris* and other notes added to provide a more precise false provenance; in Paris in the 1840s there were craftsmen skilled in forging medieval documents in the interests of family history.[5]

Unlike his father, the fifth Earl of Ashburnham was no bibliophile, and he sold his father's collections. After various negotiations the Libri manuscripts, the Palestine map among them, were sold in 1884 to the Italian government, which placed them in the Biblioteca Medicea Laurenziana in Florence; there they remain. However, manuscripts that Delisle had clearly identified as stolen from French libraries were excluded from the sale and were bought by the Bibliothèque nationale.[6]

Did Libri and his craftsmen tamper with the map of Palestine? There can be no serious doubt that they did. When the map was first drawn the arm of the Red Sea at the right-hand end ran north and south and, like the Mediterranean at the foot of the map, it was painted blue. It has been erased and, though they are not now visible, we may be fairly confident that other details were erased along with it, details that appear

5 The preface to Léopold Delisle, *Bibliothèque nationale: Catalogue des manuscrits des fonds Libri et Barrois* (Paris, 1888), pp. xii-xxxv, gives a full account of Libri's misdeeds and of the sale of his manuscripts.

6 Ibid., pp. xlv-lxiii.

23. South end of the map as originally drawn. A reconstruction of its likely appearance before the mid-19th-century alterations.

on and around the redrawn Red Sea: its own name, *Mare rubrum*, a note of the crossing of the Israelites, the town at its head and the name of the adjacent mountains, *Mons syna qui et oreb* **[23]**. A new Red Sea has been drawn, running diagonally north-east and south-west, a position unknown to medieval cartography, and the lower mountain range has been extended over the end of the original Red Sea.[7] On and around the new Red Sea, painted blue overlaid with red, the erased details have been reinstated, including a new town symbol named *Helim*, possibly for Elath; the hand of these inscriptions is not the hand of the rest of the map, but is very similar and may be seen as a deliberate imitation **[24]**.

7 It probably originally ended at about the fold and crack in the parchment. The green of the range south of this point is of a slightly different shade.

24. Hands of the original mapmaker (below) and of Libri's craftsman (above). The differences are minute but consistent in, for instance, the spacing of the letters and the form of the letter a. Florence, Biblioteca Medicea Laurenziana, Ashburnham Libri ms 1882.

The purpose of these changes is clear. The shape of the redrawn Red Sea is blatantly phallic, and it points towards the map's upper range of mountains. On other maps these are named the mountains of Gilead. Here, in the hand of the inscriptions around the new Red Sea, they are called *Montes Isaraelitarum*, the mountains of the Israelite women; the first *a* has a mark above it, and it has been read as a *d*,[8] but either way it is an inexpert misspelling. By *montes* we should understand *montes veneris*, so that what we have is a scarcely veiled erotic drawing. Beside the mountains lies the Dead Sea, named the lake of Sodom (*Lacus Sodome*); this name is in the hand of the original mapmaker, not of Libri's craftsman, but we may be sure that Libri saw it as part of his picture, and it may indeed have been what gave him the idea of making these alterations.[9] They are to be seen not as mere schoolboy obscenities but as an attempt to enhance the map's sale value. There would be private collectors prepared to pay a premium for what would have been sold to them as erotic fantasy emanating from a monastic scriptorium.

Having eliminated Libri's alterations, we are left with an impressive and important map. It is slightly larger than Madden's estimate: 29 inches from side to side, 10½ from top to bottom. With the Red Sea on the right, east is at the top and the Mediterranean coast at the foot. The map is focused on the river Jordan, formed from the union of the Jor and the Dan, but it extends not only to the Red Sea in the south, but in the north to Damascus and the rivers Abana and Pharpar beyond the mountains of Lebanon. The mountains beyond the Jordan close the map to the east; beyond them lies simply Arabia, named with widely spaced letters. The Jordan flows to the Dead Sea not only through the sea of Galilee but also, further north, through the lake, here *Malahac*, that reappears as *Aque Maron* on the large and small Burchard maps. The river Kishon is shown, and an unnamed river that reaches the sea beside Acre. Meribah, where Moses struck water from the rock, is shown as a small lake below the mountains south-east of the Dead Sea; it is named, as in the Vulgate, *aqua contradictionis*. There is another small lake, unnamed, in the mountains of Lebanon, a series of five pools appears at the south end of the sea of Galilee and the tiny Jacob's well, *puteus*, by Sychar, is in the exact centre of the map.

Except the Dead Sea, all these inland waters are painted a much brighter blue than the dull colours of the Mediterranean and the (original) Red Sea. The Dead Sea has the same bright blue around its shore but is otherwise a deep grey, while the only other water shown on the map, another small lake in the mountains east of the Dead Sea, has these colours reversed, being bright blue edged with deep grey. Mountainous areas and individual peaks are outlined in ink and painted olive-green; the colour is applied more strongly along the left side of the peaks to give a shadowed effect. Several are named: Mount Nebo, the mountains of Lebanon and a few others. Two neighbouring peaks lie outside the ranges of hills. One is named as Mount Tabor, the other, correctly Mount Gilboa, source of the Kishon, seems to have been given the name of the nearby town, Endor, while Gilboa is named on the hills further south.

Magnificently, towns are marked with golden squares or rectangles; each is outlined with a fine red line and crossed with two lines each way to make a checker pattern. There seems to be no similar symbol for towns on any other medieval map, and if it represents a stylised street pattern in plan, as our modern eyes would see it, it is of great

8 Thus Röhricht, 'Karten und Pläne VI', plate V.

9 This is an unusual name for the Dead Sea, but we find sea of Sodom in the account of the Holy Land by the Russian Abbot Daniel in about 1106-8 (Wilkinson, *Pilgrimage 1099-1185*, p. 137) and *lacus Sodome et Gomorre* in the account by Belardo d'Ascoli, written between 1112 and 1120 (*Itinera*, ed. De Sandoli, II, p. 48).

interest in the development of cartographic concepts. The lines are carefully drawn, apparently ruled, unlike the symbol at *Helim*, added by Libri's craftsman, where the lines curve to follow the shore-line of the (new) Red Sea. At Acre and Beirut, on the coast, the harbour intrudes on the rectangle and at Tyre and Sidon, where the town and harbour are on a peninsula, the rectangular shape is abandoned to cover the whole peninsula with gold.

Fourteen places are marked with a different, castle-like, symbol: a red rectangle with tower above, mostly with a cornice and either a door or two windows. Half of these are castles of the crusaders: Châteauneuf, Toron, La Fève, Kerak, Montréal and two, unnamed, between Tyre and Acre that may be seen as Scandelion and Casal Imbert. Macheronta, named north of the Dead Sea, is another castle, said to have been the castle of Herod where John the Baptist was beheaded.[10] Why the other six places – Magdala, Bosra, Bethabara, Ashdod, Emmaus and an unnamed symbol north of Sebaste – are marked with this symbol is not clear. Nor is it clear what links nine places marked with the same red symbol but without the central tower, though they may simply have been seen as small settlements, villages rather than towns: Sarepta, Cana, Sepphoris, Jenin, one unnamed that is probably Haifa or Chastel Pèlerin (another crusader castle), *Calchar'*, Nob (*Betnuba*) and Bethany, with another beside Bethany named, probably, *Via*, which may be a garbled reference to the road to Bethany where the risen Christ appeared to the apostles. A single church-like symbol appears on the mountains south-east of the Dead Sea; it marks the tomb of Aaron on Mount Hor and is the map's only feature to be painted mauve.

Some places are named without symbols: Dothan, Megiddo, Bethel, En-gedi, Modin, Shiloh and Gibeah. Further notes mark particular areas or sites, somewhat erratically. The plain of Medan appears west of the mountains of Lebanon, and the region of the people of Jerash east of the sea of Galilee. Just north of the sea of Galilee is Jacob's ford (*vadum Iacob*), where he wrestled with the angel. To the west of the sea of Galilee is 'of the five loaves' (*v panum*), where Jesus miraculously fed the five thousand, two notes on the Jordan mark the area where John baptised and the place of Jesus's baptism, and further south are the desert where the manna fell and the valley of tears where Adam and Eve were said to have mourned the death of Abel for a hundred years. Three of the twelve tribes are named in two inscriptions: 'the tribes of Zebulun and Naphtali' west of the lake *Malahac* and, simply 'Dan', almost lost in the olive-green of the mountains north-west of Jerusalem. Idumea, Samaria and Judea are each named in large letters across the region, but the neighbouring lands of Arabia and Egypt are modestly named in lower-case style at the south end of the map. However, the name of Arabia is repeated along the top of the map in large letters used also for Syria in the top left corner.

Madden dated the map to the twelfth century in his diary, and it was probably Madden who gave the map not only the number but also the same date that is ascribed to it in later catalogues.[11] It cannot be earlier than the construction of Kerak in 1142, while the castle-like symbol at Bosra may reflect the crusaders' expedition against it in 1147 and that at Ashdod the small castle built there in 1153. In fact its handwriting points clearly to the late twelfth century. Surprisingly, however, when Léopold Delisle saw it at Florence in about 1885 he described it as 'drawn in the 14th or 15th

10 *Josephus*, ed. by H.St J. Thackeray et al., Loeb Classical Library (10 vols, London and Cambridge Mass., 1926-65), IX, p. 85.

11 Thus *A catalogue of the manuscripts at Ashburnham Place* (London, 1853); *Eighth report of the Royal Commission on Historical Manuscripts* (London, 1881), p. 71. However, in Tesi, *Monumenti*, p. 25, it is assigned to the 13th century.

century, probably by an Italian hand'.[12] One does not lightly question the opinion of so distinguished a palaeographer, but in this instance he was mistaken. He was on a visit to the Biblioteca Medicea Laurenziana at the invitation of the librarian to examine the manuscripts that had once been Libri's, and if he saw them in numerical order he will have seen the map near the end of what must have been quite a gruelling ten days.[13] He will reasonably have assumed that the map had been assigned to the twelfth century by Libri, of whom he had no opinion; had he thought that the date was Madden's he would certainly have considered it more seriously.

The hand of the map is in fact remarkably like that of the maps of Palestine and Asia that were drawn at Tournai also in the late twelfth century (chapter 6, p. 41) **[25]**. The Ashburnham Libri map is not by the same writer, but it is reasonable to suppose that it too comes from north-east France or Flanders. This would accord well with its likely immediate sources, and we can see how Libri may well have acquired it while inspecting provincial libraries. In passing we may note that Tournai, in Belgium, lay outside his terms of reference. While Delisle suggests an Italian origin we should see this in the context of the later date that he proposed. Reinhold Röhricht, who accepted the twelfth-century date, also suggested that the map came from Italy because of its spelling *Scalon* for Ascalon; however, Ascalon was sometimes written without the initial *A* in north-west Europe as well as in Italy in the twelfth and thirteenth centuries.[14]

The map is drawn on a single sheet of parchment and the curve in the lower left side shows that this was the neck end. The parchment has been cropped since the map was drawn; the edges of the map must have been further away from the names of Syria in the top left corner and Arabia along the top when they were written, and we may suppose that about a centimetre, probably not more, has been cut from the top of the map and from its left side.[15] At some point the parchment was folded three times, at 18, 36 and 54 centimetres from the left end. The folds have produced some rubbing, and at the right-hand fold a split has developed in the parchment. However, it is probably because the map was long kept folded that its colours are so bright and its general condition so good. The folded map was perhaps placed in the bound manuscript of some relevant text; a very few pinholes are visible along the top and bottom, but no sewing holes, and if kept loose in the volume it would have presented Libri with an open invitation to remove it. The map may have been cut down to fit the volume or, of course, by Libri to remove some incriminating evidence of provenance. It is now held on a mount, framed and glazed; no part of the map is masked, but the dorse is unfortunately inaccessible, so that any written note or any

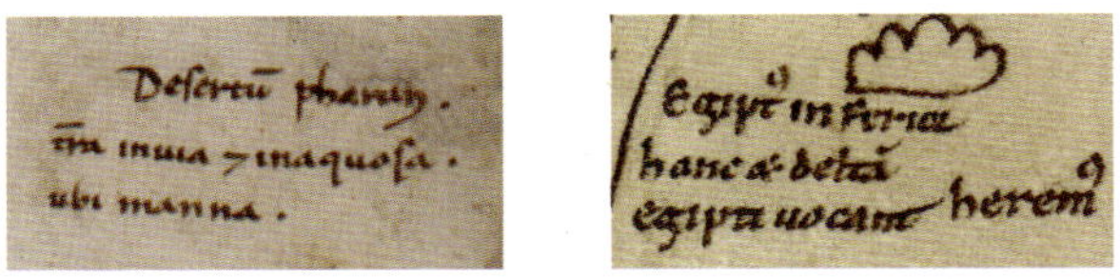

25. Hands of the Ashburnham Libri map (left) and of the Tournai maps (right). While they are not by the same writer, the hands are alike in the forms of individual letters and in general style. Florence, Biblioteca Medicea Laurenziana, Ashburnham Libri ms 1882; BL, Add. MS 10049, f. 64v.

12 'Le catalogue annonce . . . une carte coloriée de la Palestine, du XIIe siècle. Il ne s'y trouve qu'une carte assez grossière de la Terre sainte, tracée au XIVe ou au XVe siècle, probablement par une main italienne, sur une feuille de parchemin haute de 750 millimetres et large de 280' (Léopold Delisle, 'Notice sur des manuscrits des fonds Libri conservés à la Laurentienne', *Notices et extraits des manuscrits de la Bibliothèque nationale et autres bibliothèques*, 32 (1886), p. 97).

13 Ibid., pp.1-3.

14 Röhricht, 'Karten und Pläne VI', p. 177. In the texts printed in *Itinera*, ed. De Sandoli, Scalon, with variants, occurs on I, p. 318 (Caffara di Caschifelone); III, pp. 38 (Fifth Anonymous), 170 (Aimaro dei Corbizzi), 480 (anonymous Latin text); IV, p. 60 (anonymous French text). It occurs on an English early-13th-century plan of Jerusalem (Brussels, Bibliothèque royale de Belgique, ms IV 462, f. 8v, a reference I owe to the kindness of Dr Hanna Vorholt) and from the 14th century as a common noun in English, meaning shallot (*Oxford English dictionary*, 2nd edn, ed. by J.A. Simpson and E.S.C. Weiner (20 vols, Oxford, 1989), XIV, pp. 565-6).

15 While Delisle's measurements in about 1885 are some 15 millimetres greater in each direction than its actual size (above, note 12), we must acquit the Biblioteca Medicea Laurenziana of permitting the cropping after the map had passed into its custody. Delisle must have made simply a visual estimate, like Madden forty years before; and Madden, as we have seen, underestimated the size of the map.

pattern of pinholes created in drafting is now invisible. Neither Madden nor Delisle mentions the frame and it may well date from after their time; the backboard is certainly recent.

Apart from the ninth-century monastic plan at St Gall, this is Europe's oldest surviving sheet map, that is, one drawn on a loose piece of parchment, not on a page in a bound volume. It is surprising that so little notice has been taken of it; there are only two published accounts, and both are brief. In 1895 Röhricht discussed it in a single paragraph, comparing it unfavourably with the large Burchard map and its derivatives, but excusing it because of its earlier date. Along with this he published a line drawing of the map with its inscriptions typeset.[16] This was the only reproduction of the map until 1992, and it was far from adequate; no sign of the original Red Sea appears and a few of the inscriptions have been misread or omitted.[17] In 1992 the map was included in a major exhibition at Genoa commemorating the first voyage of Christopher Columbus, and in the published catalogue it is reproduced in full colour. The accompanying text by Alberto Capacci, more than a page, discusses medieval maps of the Holy Land in general, then describes the present map.[18]

The map is an obvious fair copy and it was probably the work of a single hand at a single operation. It offers practically no evidence of how the mapmaker worked. No lines or other marks of drafting are visible and we can only await examination of the dorse to discover any pattern of pinpricks used in its construction. The outlines of the hills were drawn before the symbols for particular places (at Bethany the outline is visible below the red of the symbol) and the hills were coloured olive-green before the gold and red were added to mark the towns, castles and other places, and also before the grey surround was added to the lake east of the Dead Sea. Here, and in the Dead Sea itself, the blue was applied before the grey. The inscriptions were probably written as the last stage in making the map. An odd feature of their hand is a faint stroke, like an acute accent, that appears for no obvious reason above some words of the inscriptions.[19]

There are some indications that the map was copied not from an entirely new draft but from an existing map. The confusion that led to Mount Gilboa being given the name of the town Endor can have come about only through misunderstanding a map and the names of Arabia and Egypt in small letters at the south end of the map point the same way; *Arabia* duplicates the name along the top of the map and one would anyway expect both to be in large letters like the map's other regional names. It looks very much as if they were late additions to the exemplar, picked up and incorporated in this copy.

A world map presumably provided the map's framework: coastlines, rivers, mountains and principal towns. While no surviving map of the world or of the region reproduces the same overall pattern, individual features can be paralleled on other maps. We see the same straight Mediterranean coast on the Second Tournai map of Palestine and on Matthew Paris's Oxford map, and its few gentle indentations recall those on the Ebstorf map. Tyre is shown (rightly) as a peninsula on several other maps besides this, while on the Ebstorf map Sidon too, as here, protrudes into the sea. The two rivers that flow into the Mediterranean reappear on Matthew Paris's Oxford map, while on the Hereford map the Dead Sea lies, surprisingly, east of the main axis

16 Röhricht, 'Karten und Pläne VI', pp. 177-8, plate V.

17 E.g. *IVDEA* appears simply as *DEN*, and *Olivet* is left out altogether. It is unfortunate that Röhricht 'Karten und Pläne VI', p. 178, points to the name *Aron* for Toron as a mistake of the mapmaker, for the mistake is his own; the map actually reads *Tiron*.

18 *Cristoforo Colombo*, ed. Cavallo, pp. 461, 464-6. Regretfully, I cannot agree with Alberto Capacci's suggestion there that the symbols for places can be related to notes of towns and villages on Matthew Paris's Oxford map; see chapter 7, note 25.

19 E.g. *Bethabra*, *Endor*, *Gerasa*, *Mons Or*, *Naym*, *Sapheth*.

of the Jordan, just as it does here. The large Burchard map also shows mountain areas in outline, though in a rather different pattern, and as we have seen, the lake between the sea of Galilee and the sources of the Jordan recurs on both the large and small Burchard maps. However, the Ashburnham Libri map is unique in showing the Red Sea to the south of Palestine – on all other maps it is to the east or south-east – and on no other map do the mountains of Lebanon run so uncompromisingly north and south.

Besides a map of the world, two other sources contributed to the Ashburnham Libri map. One was a written account of the Holy Land that must have belonged to the closely related group of twelfth-century texts that includes those of Rorgo Fretellus and Theoderic and the so-called Sixth Anonymous text. Some places mentioned in all three of these texts are named on this map but on no other, such as Carnaim and the tomb of Aaron, while others named also on the Burchard-based maps must here come from the twelfth-century text, among them Kedar, Macheronta, Jenin and Dothan.[20] All three texts call attention to the tribes of Naphtali and Zebulun in Galilee; Fretellus and the Sixth Anonymous describe the valley of tears and the plain of Medan; Fretellus tells how the river Pharpar enters the sea at St Symeon, which would explain why here, as on no other map, it is shown flowing due north, to turn west beyond the map's northern border.[21] However, none of these texts mentions the lake *Malahac* to the north of the sea of Galilee, and their name for the Dead Sea is the Asphalt Lake (*Lacus Asphaltidis, Asphaltitis*), not the map's reasonable, but most unusual, lake of Sodom which, we have seen, may have provided the inspiration for Libri's alterations.[22]

The further additional source that contributed to the map, that is, our map's exemplar, may have been a written text not yet identified or lost altogether, but it could well have been the personal knowledge of a returning crusader. Certainly the published texts took some account of current affairs, and all three mention the castle of Montréal built by the crusader King Baldwin I in 1115.[23] But the map records other crusader castles that appear in no published twelfth-century text of this sort and on no known world map: Toron, Châteauneuf, La Fève and others. The map is an eloquent picture of the Holy Land while much of it was in Christian hands and it was a map of some beauty. It is sad that it should have fallen victim to the pathetic vandalism of Guglielmo Libri.

20 *Rorgo Fretellus*, ed. Boeren, pp. 15, 16, 23, 24, 27; *Peregrinationes*, ed. Huygens, pp. 178, 182, 188, 189, 191; *Itinera*, ed. De Sandoli, III, pp. 52, 60, 62, 64.

21 *Rorgo Fretellus*, ed. Boeren, pp. 9, 21, 22, 24; *Peregrinationes*, ed. Huygens, p. 192; *Itinera*, ed. De Sandoli, III, pp. 46, 58, 60.

22 *Rorgo Fretellus*, ed. Boeren, p. 11; *Peregrinationes*, ed. Huygens, pp. 181, 190; *Itinera*, ed. De Sandoli, III, p. 46.

23 *Rorgo Fretellus*, ed. Boeren, p. 18; *Peregrinationes*, ed. Huygens, p. 178; *Itinera*, ed. De Sandoli, III, p. 54.

6 THE TOURNAI MAPS
late 12th century[1]

London, British Library, Add. MS 10049, f. 64r,v. Parchment, 35 × 23 cm.

Recto: first map of Palestine **[26]**, landscape format, deep brown and red ink, *overlaid and replaced by* map of Asia **[27]**, portrait format, deep brown and red ink with red paint.

Verso: second map of Palestine **[28]**, landscape format, deep brown ink with yellow paint, *partly overlaid and replaced by* third map of Palestine **[29]**, landscape format, deep brown ink.

THE TOURNAI MAPS, misleadingly known as the Jerome maps, are on the last leaf of a late-twelfth-century manuscript containing two original works by St Jerome (died 420) and his Latin translation of the *Onomasticon* written in Greek by Eusebius, *Liber de situ et nominibus locorum hebraicorum*, known for short as *Liber de locis*.[2] All three interpret or discuss the names of persons and places and other words that occur in the Bible; each is arranged alphabetically but by initial letter only. The maps are a natural companion to at least the *Liber de locis*, but are not directly linked to the text; few of the places listed are named on the maps. On the front of this last leaf is a map of Asia in upright, portrait, format, with east at the top and with Asia Minor, Anatolia, drawn disproportionately large **[27]**. On the back, drawn lengthways, in landscape format, is a map of Palestine, again with east at the top; Jerusalem, prominent but not so large as to dominate the map, is drawn as a distinctive double circle with at one side the spire-capped tower of David on Mount Sion **[29]**. The leaf with the maps can be seen as an integral part of the manuscript, not as an insertion; it almost certainly forms a bifolium with the first leaf of the volume's last quire.[3]

The book belonged to St Martin's Benedictine abbey at Tournai, now in Belgium: on the first page, in a late-twelfth-century hand, is 'Book of St Martin of Tournai. Whoever takes it away or allows it to be taken away shall be anathema. Amen', and there are similar notes on two other pages.[4] The abbey numbered each of its books with a roman numeral written on the last or penultimate page or pastedown, and here we have indeed the pressmark 'lx' at the top of the last page, the map of Palestine.[5] It is just possible that it is the book listed in the abbey's library catalogue, drawn up not long after 1159, as 'Jerome on the Hebrew names in one volume'.[6] Certainly, however, we need not doubt that St Martin's abbey at Tournai was where the texts were copied and the maps drawn.

An account of the abbey in the early twelfth century names some of the monks who copied books and gives an interesting glimpse of its scriptorium in the time of Abbot Odo (1095-1105): 'He delighted in the number of scribes that the Lord had given him, so that if you went into the cloister you would commonly see twelve young

1 I am grateful to both Prof. Patrick Gautier Dalché and Dr Stéphane Lebreton for helpful discussion of the problems arising from the Tournai maps and of the important and highly relevant work that each has done. A collaborative volume focused on the maps, edited by Dr Lebreton, is in preparation; it will examine in greater detail some of the questions discussed here.

2 *Liber hebraicarum questionum in Genesim*, ff. 1v-21r; *Liber de nominibus hebraicis* (here named *Liber nominum* or *Liber de interpretationibus nominum veteris et novi testamenti*), ff. 22v-43v; *Liber de locis*, ff. 44r-63v. The first two, edited by Paul de Lagarde, are in *S. Hieronymi presbyteri opera*, I: *Opera exegetica*, vol. 1, Corpus christianorum, series latina 72 (Turnhout, 1959), pp. lxi-lxv, 1-161; the third is in *Eusebius Werke*, III, part 1: *Das Onomastikon der biblischen Ortsnamen*, ed. by Erich Klostermann (Leipzig, 1904). An earlier edition of each of the three is reprinted in *Patrologiae cursus completus, series latina*, ed. by J.-P. Migne, 23 (Paris, 1845), respectively cols 935-1010, 771-858, 859-928.

3 Although the quires are thrown out on guards, so that collation is straightforward, the join of the first and last leaves of the final quire (ff. 60, 64; the second leaf of the six-leaf quire is cancelled) has been so heavily repaired that we can be only 'almost' certain that they are a single piece of parchment. But the maps' history points strongly in the same direction.

4 'Liber sancti martini tornacensis. Quique abstulerit uel auferri permiserit sit anathema. Amen', f. 1r; similarly on ff. 1v, 63v.

5 *Corpus catalogorum Belgii: the medieval booklists of the southern Low Countries*, ed. by Albert Derolez et al., IV: *Provinces of Brabant and Hainault* (Brussels, 2001), p. 318.

6 'Iheronimus de hebraicis questionibus in I volumine', ibid., p. 323.

monks seated in chairs and writing in silence at carefully and skilfully made desks.'[7] It need not have been so very different when our manuscript was written, some generations later. Its three texts are written in a single hand, but it was another hand that produced the two maps and we see this hand of the mapmaker in other manuscripts from Tournai. In one he was responsible for a rudimentary diagram in the margin beside Bede's text on the holy sites, *De locis sanctis*.[8] He was perhaps the scriptorium's graphics man. But interestingly he entered other marginal notes, mostly of place-names, in a copy of Antoninus of Piacenza's *Tour of the holy places* (*Perambulatio locorum sanctorum*) **[30]** and himself copied a short text, with other extracts, on the geography of the Holy Land.[9] It looks as if his interests came together in producing the maps.

One remarkable feature of the Tournai maps is an oval hole in the parchment, about 4 by 2.5 centimetres, that has been covered with a parchment patch, neatly sewn in place with a thin strip of parchment around the edge. Ingeniously, the mapmaker has used this patch as the island of Crete on the Asia map, with a note that Crete, with the Cyclades, was the seventh province of the Greeks and had a hundred cites. On the other side, on the map of Palestine, the patch has a range of mountains drawn on it, the Caucasus. We shall see that this parchment page has suffered many erasures, but it was not these that produced the hole; it is a natural lacuna. There are similar lacunae on eleven of the other sixty-three leaves in the manuscript, while other leaves have been cut from the edge of skins, with wide curved gaps at the side, or have had tears repaired by sewing apparently early in their history.[10] We find just the same pattern in other Tournai manuscripts of this time,[11] and we may suppose that the scriptorium's output demanded careful economy in the purchase of parchment; it was bought as cheaply as possible and never squandered.[12] This is an important point that does much to explain the complicated story of the maps' production.

St Martin's abbey was dissolved in 1796 and its library was dispersed. The volume with the maps at some point belonged to Amans-Alexis Monteil, the French historian (1769-1850); he is named – 'Manuscrit app[t] à Mr Monteil' – in the title stamped on the preserved front cover of the volume's former binding.[13] Clearly, however, it had passed from his hands before the sale of his manuscripts in 1835, for in 1836 it was included in a sale of manuscripts that had belonged to the English collector Richard Heber (1773-1833).[14] In its description of the manuscript the sale catalogue gives us our earliest reference to the maps: 'At the end is a remarkably curious and old map of Egypt and the Holy Land.' It was bought by the British Museum for £9.5s. and numbered Add. MS 10049.[15] The maps were entered briefly in the Museum's catalogue of manuscript maps in 1844, they were described in 1850 by Vicomte de Santarem, who also published a redrawn copy of the map of Asia, and were briefly mentioned in 1873 in the book on the Hereford world map by Bevan and Phillott, who wrote that the two maps 'together are well nigh tantamount to a map of the world'.[16] It was only in 1895 that Konrad Miller published the first – and still the only – detailed account of the two maps, with collotype and redrawn reproductions of each and with a full transcription and discussion of all their inscriptions.[17]

Miller argued strongly that these are our only copies of maps that Jerome drew himself, adapting them from maps drawn by Eusebius. He took the Tournai

7 'si claustrum ingredereris, videres plerumque 12 monachos iuvenes in cathedras sedentes et super tabulas diligenter et artificiose compositas cum silentio scribentes', in the continuation of Herrimann's 'Liber de restauratione monasterii S. Martini Tornacensis', *Monumenta Germaniae historica: scriptores*, 14 (Hanover, 1883), p. 313.

8 BL Add. MS 15219, f. 14r.

9 BL Add. MS 15219, ff. 3r-11r, 2r,v. The incipit of the text (f. 2v) is 'Ierusalem est in qua regnauit adonibezec'.

10 Lacunae on ff. 10, 14, 15, 17-19, 38, 42, 44, 53, 62; curved sides on ff. 27, 30, etc.; repaired tears on ff. 33, 53, 55.

11 Thus it appears in BL Add. MSS 15219, ff. 17, 47, etc.; 25312, ff. 9, 10, etc.; 38840, ff. 5, 37, etc.

12 James E. Thorold Rogers, *A history of agriculture and prices in England* (7 vols, Oxford, 1866-1902), I, pp. 643-4; II, pp. 570, 574, 576; and M.T. Clanchy, *From memory to written record: England 1066-1307* (London, 1979), pp. 92-5, both stress the cheapness of parchment in 13th- and 14th-century England, but it may well have cost more in 12th-century Tournai and it is not wholly irrelevant to say that we would today consider a quarter or a third of the cost of a day's unskilled labour a lot to pay for a sheet of paper.

13 There is a short account of his life by 'A.J.' in *Nouv. biogr. gén.*, XXXVI, cols 169-70.

14 There is an account of Heber by Arthur Sherbo in the *ODNB*, XXVI, pp. 214-16.

15 Robert Harding Evans, *Bibliotheca heberiana: catalogue of the library of the late Richard Heber, esq. Part the eleventh. Manuscripts* (London, 1836), p. 86 (lot 831); interleaved and annotated copy in the British Library. In his diary Sir Frederic Madden records buying manuscripts for the British Museum at this sale, 10-20 February, but does not specifically mention this volume (Oxford, Bodleian Lib, MS Eng.hist c 151, pp. 79-88).

16 *Catalogue of manuscript maps, British Museum*, I, p. 13; Vicomte de Santarem, *Essai sur l'histoire de la cosmographie et de la cartographie pendant le Moyen Âge* (3 vols, Paris, 1849-52), II, pp. 245-7; Vicomte de Santarem, *Atlas composé de mappemondes, de portolans, et de cartes hydrographiques et historiques depuis le VIe jusqu'au XVIIe siècle* (Paris, 1842-53), map 103; Bevan and Phillott, *Mediaeval geography*, p. xlvi.

17 Miller, *Mappaemundi*, II, plates 11, 12; III, pp. 1-21, plate I.

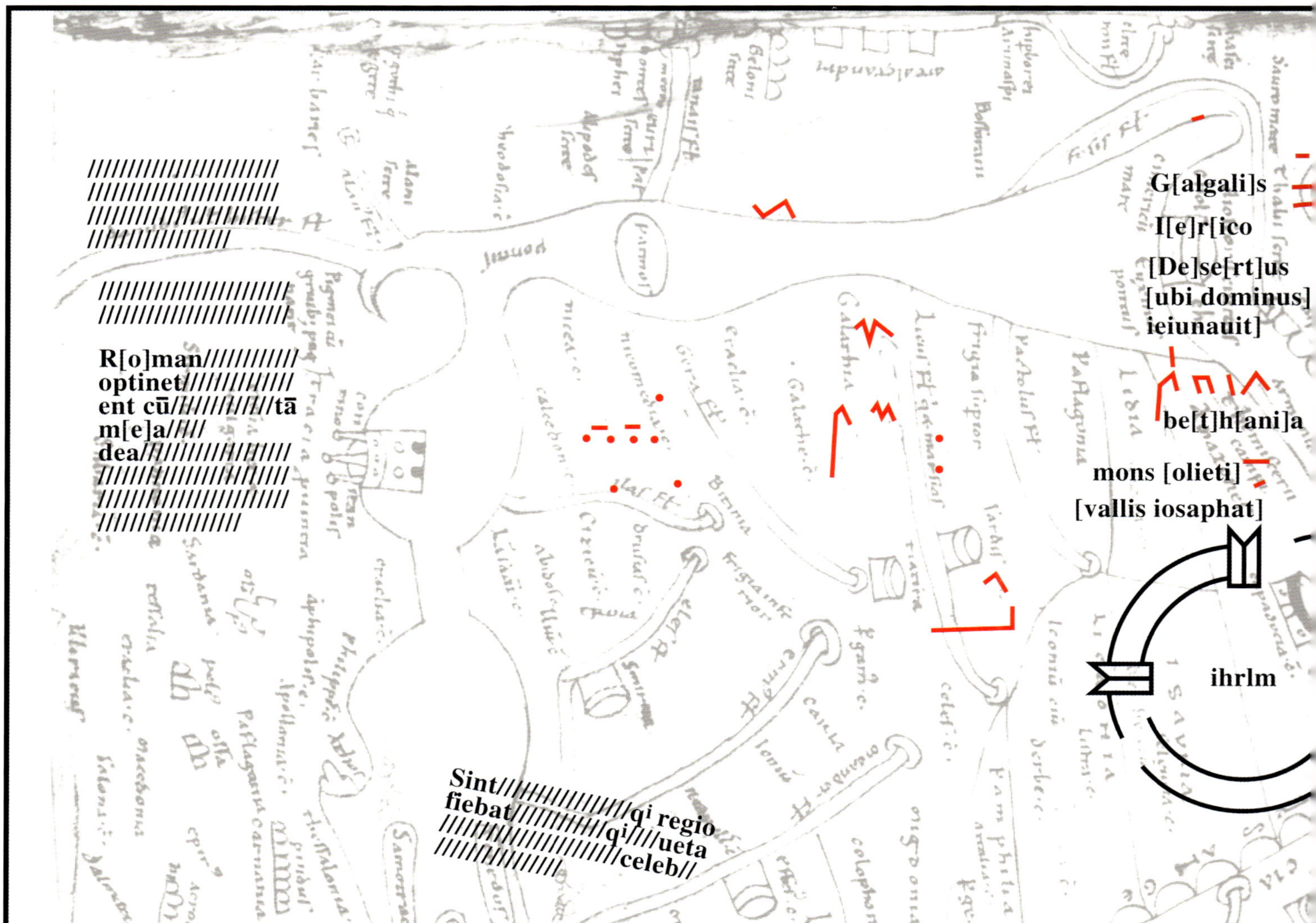

26. The first Tournai map of Palestine, late 12th century; east at the top. The map was erased to make way for the map of Asia on the same page and this drawn copy is of all that can be discerned. This includes the circle of Jerusalem and the tower of David, and overall enough is visible to reveal the form and character of the map.

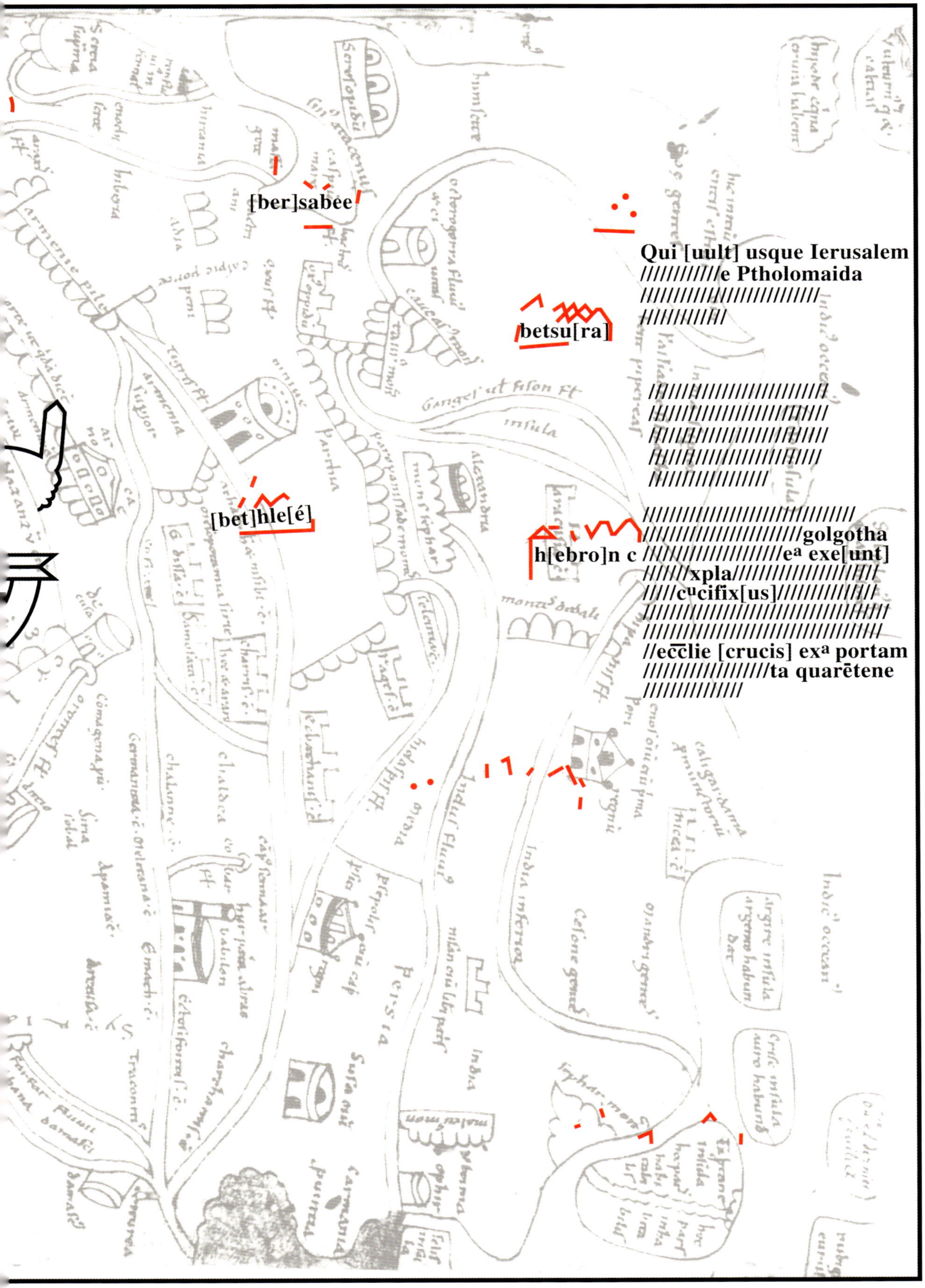
[ber]sabée
betsu[ra]
Qui [uult] usque Ierusalem
///////////e Ptholomaida
[bet]hle[é]
h[ebro]n c
golgotha
eª exe[unt]
xpla
cucifix[us]
//eccl̄ie [crucis] exª portam
ta quarētene

27. The Tournai map of Asia, late 12th century; east at the top. Anatolia, divided into provinces, dominates the lower half of the map; at the top is the Indian Ocean. Bottom right is the parchment patch used as the map of Crete. BL, Add. MS 10049, f. 64r.

pressmark, 'lx', to be a folio number copied from an exemplar, seeing it as evidence that the maps were copies; they were not the original work of the Tournai mapmaker. He points too to a reference to an appended *mappa* in a note at the end of the *Liber de locis* that Jerome translated from Eusebius. In the preface to his translation Jerome tells us that Eusebius produced a picture (*pictura*) of Jerusalem and the Temple in a chorography (*chorografia*) of Judea, which might mean either a written description or a map, and that he also defined the lands allotted to each of the twelve tribes, which appear – all or some – on certain medieval world maps. From this Miller suggests that this lost work of Eusebius, dating from 324 or 325, included several maps and that it was from them that Jerome drew the maps of Asia and Palestine. On these maps, Miller points out, there is only a single name later than Jerome's time, 'This is Moesia and Bulgaria' (*Mesia hec et uulgaria*); the Bulgars settled in Moesia only from 678. Crete as the seventh Greek province and the provinces of Anatolia shown on the map of Asia date from the fourth century, and features he said were typical of later, medieval, maps – Jerusalem at the centre and the earthly paradise – are lacking. There is some correspondence between the *Liber de locis* and the names on, especially, the map of Palestine, and there are signs that Bede (672 or 673-735) had a version of the map before him when he wrote his little book *On the place-names in the Acts of the Apostles* (*De nominibus locorum quae leguntur in libro Actuum Apostolorum*): he describes islands and provinces as if from a map, he incorrectly places Lydda, otherwise Diospolis, on the coast as it is on the map of Palestine (which in fact makes them two different places), he calls the river Cydnus (i.e. the Berdan or Tarsus River) the Cygnus, like the map of Asia, and he identifies Neapolis incorrectly with the town in Anatolia, which is shown on the map of Asia, instead of with its namesake in Macedonia, which is not.[18]

All this adds up to a persuasive argument, and it has persuaded many scholars since 1895; C. Raymond Beazley presented it in some detail to English speakers in 1900.[19] However, the evidence it deploys is entirely circumstantial. Certainly Miller's argument tells us some interesting and important things: that the maps' wording comes from the late Roman period, not from the twelfth century, and that Bede saw a closely related text – or, just possibly, a map – when he wrote his work on the Acts of the Apostles. But what it does not tell us is that Jerome drew the maps. There is not a shred of hard evidence that he drew a map of any kind, let alone the archetypes of what Miller called the Jerome maps. The note of an appended *mappa* was written not by Jerome but by the editor of his collected works in 1735 and refers to a modern engraved map.[20] We know only that Eusebius produced a representation of Jerusalem and the temple; all else is conjecture. Indeed, if Jerome had drawn maps to accompany his translation of Eusebius' work, it is most improbable that they would appear in just this one manuscript; the text survives in more than a hundred medieval copies.[21]

In fact the construction of the Tournai maps is not so straightforward as Miller supposed. Beazley in 1900 noted that 'Certain almost invisible marks show . . . that the scribe wrote rough copies of various names as if to get his hand in for the fair draught', and in 1901 he mentioned traces of red on the map of Asia, with 'a picture of a house, or temple, probably representing a city' and blocks of text, all erased, as showing that the map had been copied, not created, at Tournai.[22] These traces, these marks, are more significant than Beazley realised. Both maps are palimpsests: the present maps are drawn over earlier maps that have been more or less fully erased.

Looking carefully first at the verso of the leaf, the map of Palestine **[29]**, we see that much more has been rewritten or redrawn than 'various names'. Most strikingly the Mediterranean coast has been completely redrawn, along with the coastal towns; it

18 Ibid., III, pp. 2-4.

19 Beazley, 'New light', pp. 138-40. Here (p. 138) he wrote that 'The Jerome maps . . . were almost certainly drawn under the direction, if not by the hand, of that Father', but by the following year, when the second volume of his *Dawn of modern geography* was published, he was less confident: 'the so-called 'Jerome' maps . . . were possibly drawn under the direction, if not by the hand, of that Father' (p. 566).

20 *Sancti Eusebii Hieronymi stridonensis presbyteri opera*, ed. by Dominicus Vallarsius and Scipio Maffeius (11 vols, Verona, 1734-42), III, cols 291-4. The note is repeated in the later edition used by the *Patrologiae cursus completus*, where it unfortunately appears (23, cols 927-30) as though it was part of what Jerome wrote. An earlier edition, *Sancti Eusebii Hieronymi stridonensis presbyteri opera*, ed. by Johannes Martianay (5 vols, Paris, 1699-1706), II, cols 495-6, had also added a modern map with a similar, not identical, note.

21 Bernard Lambert, *Bibliotheca hieronymiana manuscripta: la tradition manuscrite des oeuvres de saint Jérôme*, Instrumenta patristica 4 (4 vols, The Hague, 1959-72), II, pp. 23-9.

22 Beazley, 'New light', p. 139; Beazley, *Dawn*, II, p. 605.

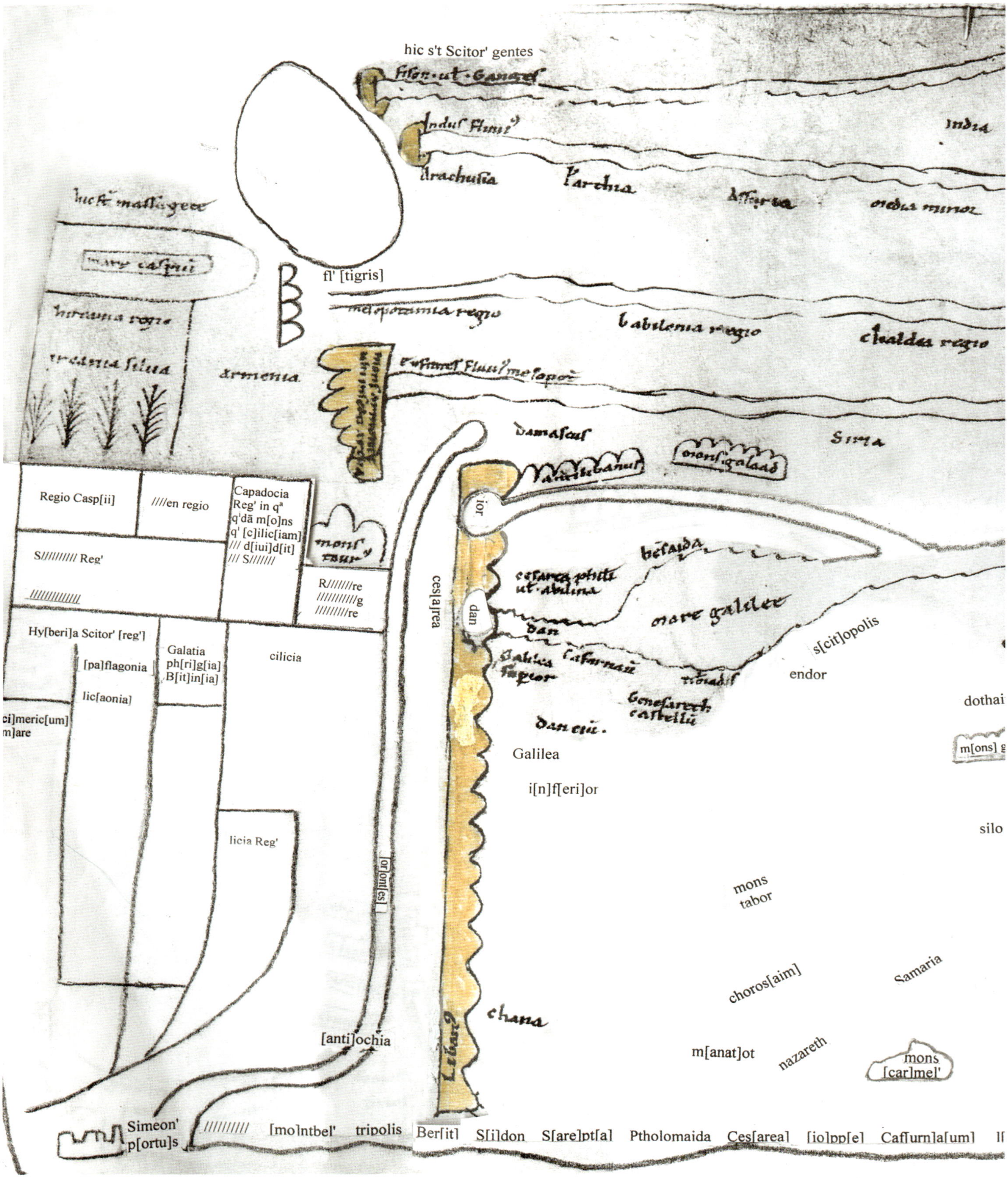

28. The second Tournai map of Palestine, late 12th century; east at the top. Most of the lower half of the map was erased and redrawn to make the third map of Palestine; not all the erasures can be read. In some cases it is uncertain whether a place-name was allowed to remain from the second map or was added in making the third.

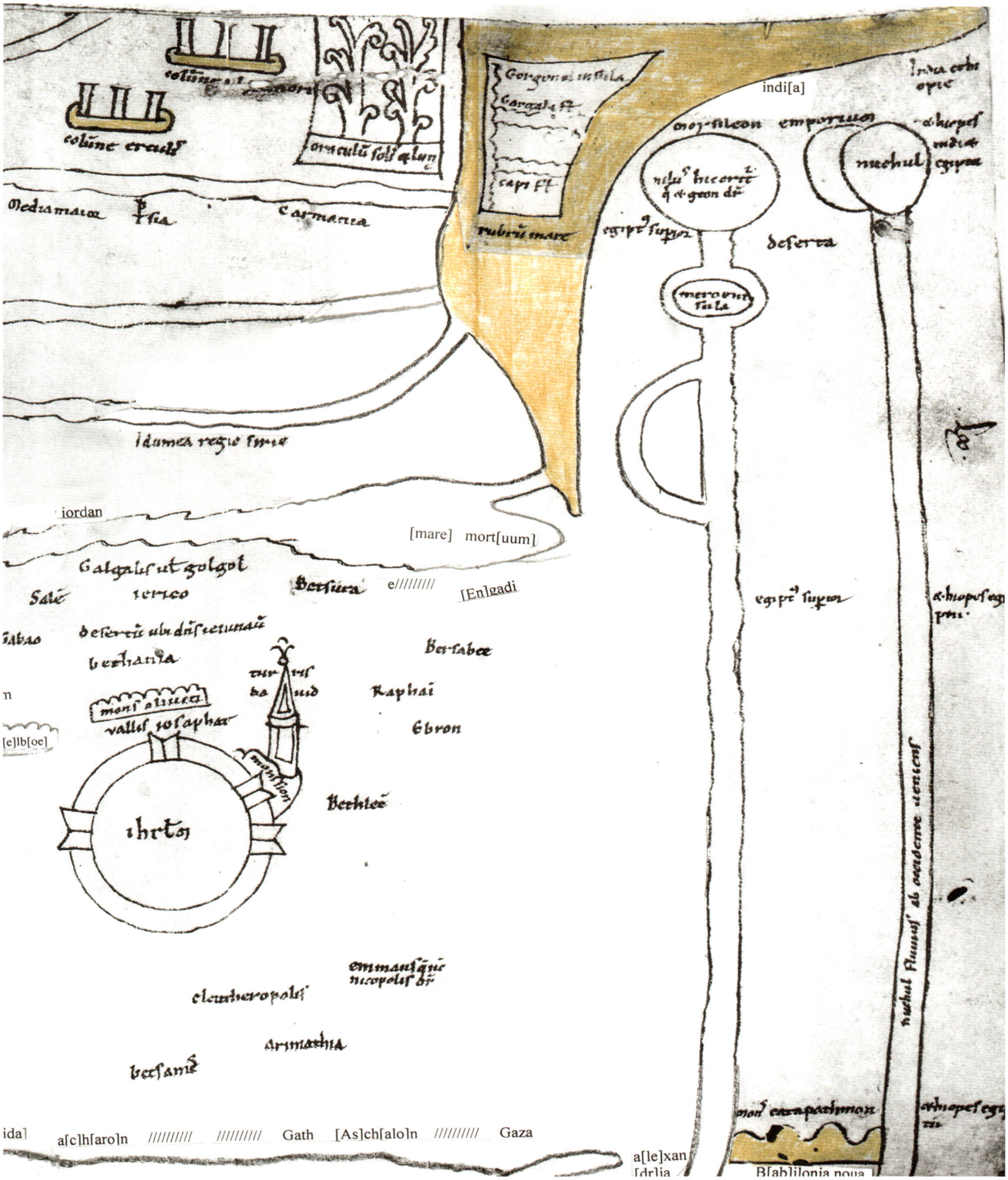

indi[a]
rubru mare
deserta
nuchul
Carmania
idumea regio
iordan
[mare] mort[uum]
Galgala ut golgot
iericho
Bersura
e/////////
[En]gadi
Bersabee
bethania
Raphai
Ebron
vallis iosaphat
[e]lb[oe]
Bethlee
ihrlm
eleutheropolis
Armathia
egipt9 super
mons carapatimon
ida]
a[c]h[aro]n
//////////
//////////
Gath
[As]ch[alo]n
//////////
Gaza
a[le]xan
[dr]ia
B[ab]ilonia noua

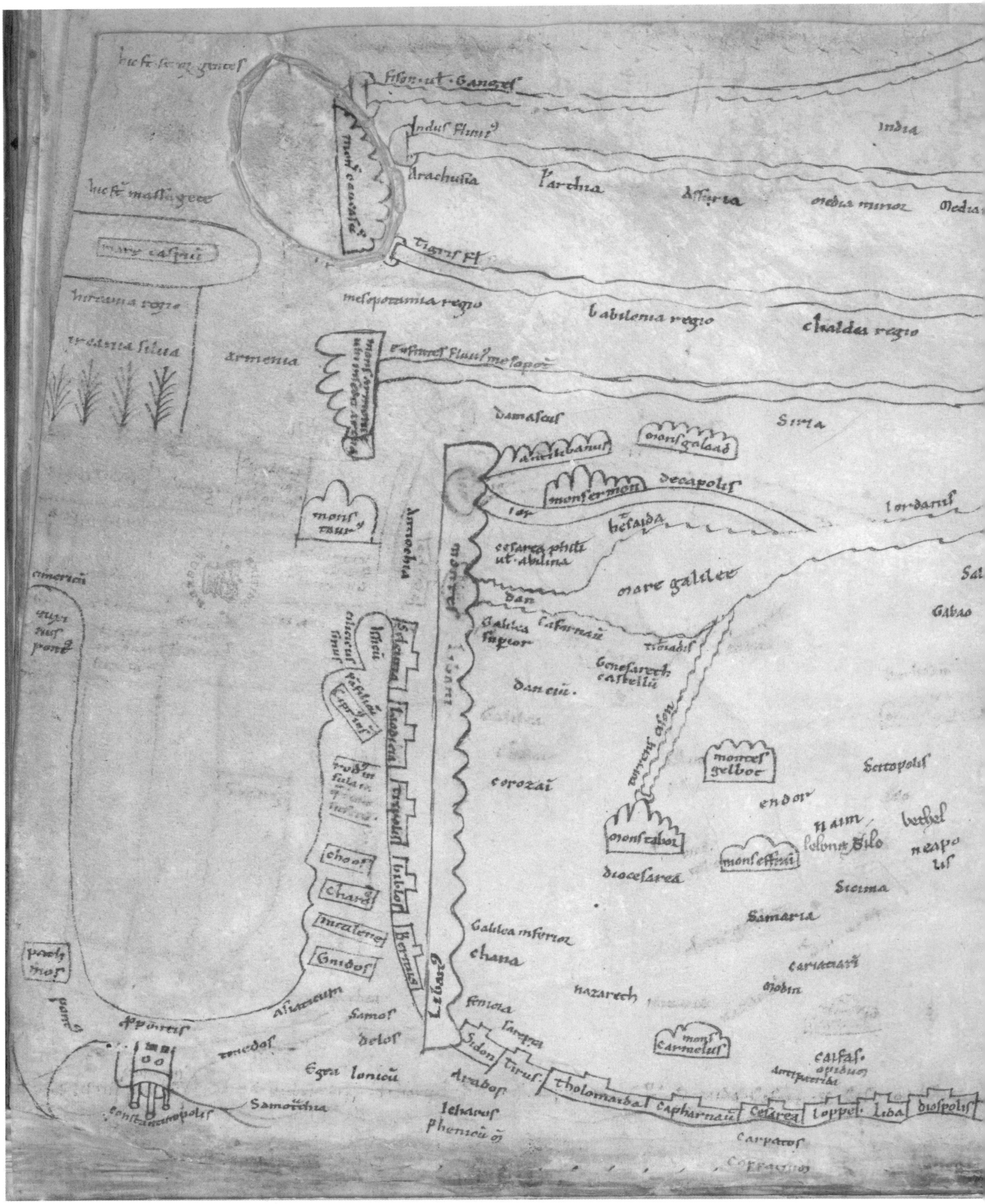

India
arachusia
Parthia
Assyria
Media minor
Tigris fl
mesopotamia regio
babilonia regio
Chaldea regio
armenia
damascus
Siria
mons galaad
mons hermon
decapolis
Iordanis
betsaida
mons tauri
mare galilee
dan
Genesareth castellu
montes gelboe
Scitopolis
endor
Naim
Silo
bethel
neapolis
mons tabor
diocesarea
Samaria
corozai
Chana
nazareth
fenicia
mons carmelus
Sidon
Tirus
Tholomaida
Capharnau
Cesarea
Ioppe
Lida
diospolis
Carpatos
Cnidos
Samos
delos
tenedos
Egea
Ionicu
Samothia
Gabao

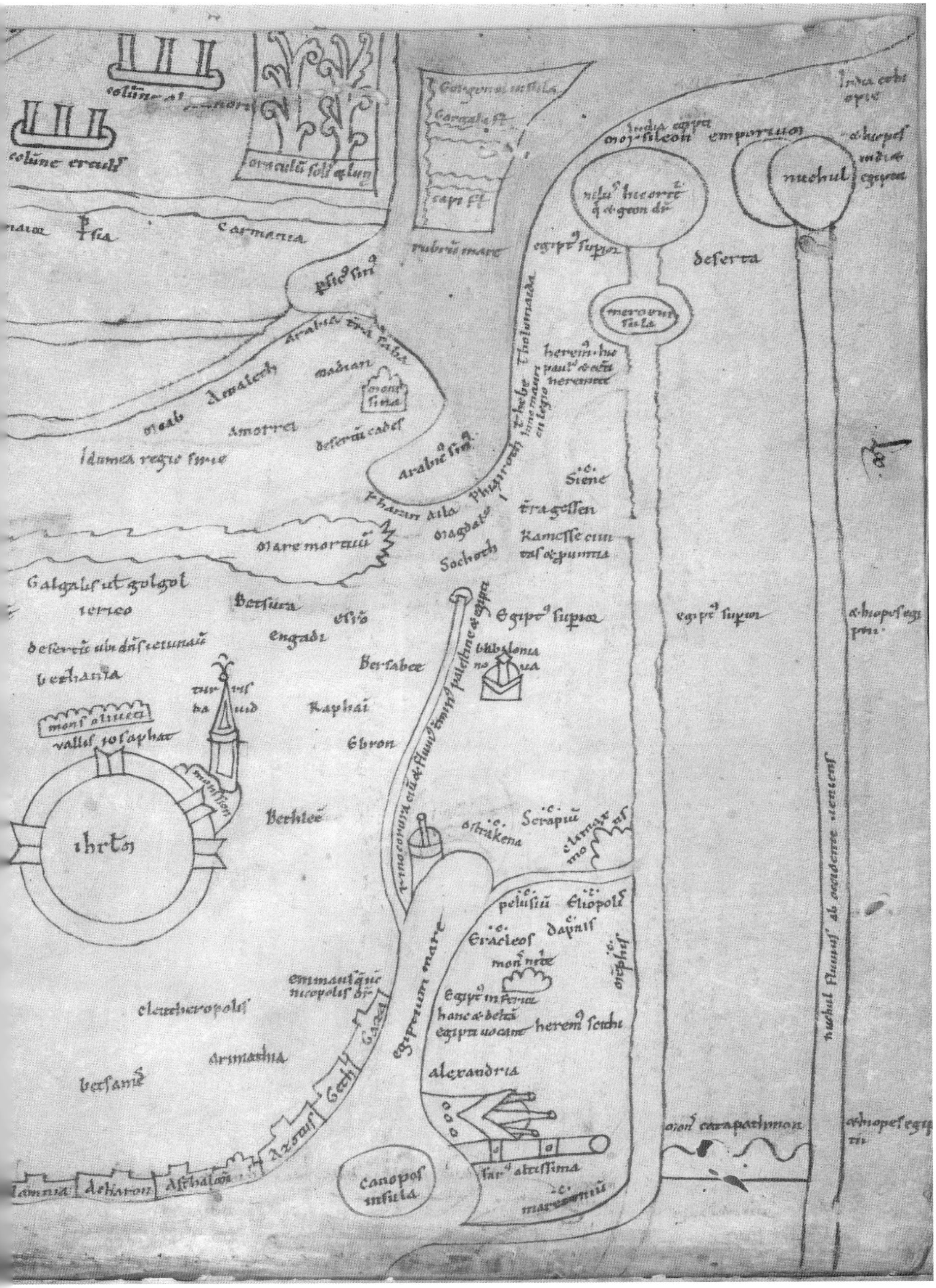

29. The third Tournai map of Palestine, late 12th century; east at the top. Especially in the upper half, much of the second map remains unaltered, but elsewhere a new coastline and many other changes entered over erasures make this a quite different map. BL, Add. MS 10049, f. 64v.

now appears with a deep bight to the north and south of Palestine, but was originally a straight unindented line. In Anatolia not only the lines of the provinces but their names and other notes have been erased, and so too, further south, has the river Orontes. There have been changes to the river Jordan, especially around the Dead Sea, and lesser changes to the courses of the Euphrates, the Tigris and the Nile. Between the Jordan and the Mediterranean names of mountains, towns and provinces have been erased and rewritten in new positions, though Jerusalem, with its distinctive shape, remains unchanged. What we see is a largely different map from what was originally drawn; we can reasonably say that one map of Palestine has been replaced by another.

Turning back to the recto of the leaf, we can see that the map of Asia has undergone some similar but much less extensive changes **[27]**. A few place-names have been erased and rewritten, and at the top of the map there have been some small changes to the coast of the Indian Ocean and to the mouths of the rivers that flow into it. These are minor alterations, not to be compared with the complete revision of the map of Palestine on the other side of the leaf. However, if we turn the map of Asia sideways, with the page's outer edge towards us, we see, close to the centre of the page, the faint but unmistakable shape of Jerusalem in the same form as on the map of Palestine. We then become aware of other erased wording written on the same axis – a few individual words and some blocks of text – as well as a few traces of red rectilinear symbols, and we realise that what was first drawn on this recto page was a map of Palestine, in landscape format as on the verso, a map that has been entirely erased so that the page could be used for the map of Asia.

We have thus not two Tournai maps, one of Asia and one of Palestine, but four: a map of Palestine on the recto that has been wholly erased and replaced by a map of Asia, to which some small changes have been made, and over the page another map of Palestine that has been so thoroughly altered as to create a new, third, map of Palestine. The map of Palestine on the recto is most likely to have been the first of the four to be drawn, and we shall call it the first map of Palestine; we cannot be certain, though, that the map on the verso in its original state, which we shall call the second map of Palestine, had not been drawn first. However, by an extraordinarily lucky chance, we can be certain that we are throughout looking at the work of a single mapmaker. We have seen how the hole in the parchment was covered by a patch, attached in fact to the recto page. The edge of the patch extends in part a couple of millimetres beyond the parchment strip that holds it in place. Wholly covered by it is the place-name *Caesarea Philippi* written on the same axis as the first Palestine map, to which it clearly belongs. The patch must have been added after the first map of Palestine had been drawn and before it was erased, but presumably with a view to its replacement with the map of Asia, for there is no sign of an attempt to replace the now obscured place-name. But, no less interesting, *Caesarea Philippi* is written in the same hand as the visible maps of Asia and Palestine, the hand of the mapmaker whose interests, we have seen, included diagrams and the Holy Land. One mapmaker was responsible for all four of the Tournai maps.

This is immensely helpful in reading and reconstructing what has been removed from the maps. On both sides of the leaf erasure has been more thorough in some parts than in others, but everywhere the erased lines and words are very faint – as is shown by the little attention that writers on the maps have paid to them. Knowing the writer's hand means that we know what letter-forms, what general characteristics of the writing, we may expect. Even so, what we mostly have are the merest shadows of what was drawn and written, and these shadows can be interpreted only with great difficulty. The readings offered here should be seen as no more than provisional. Different eyes and newer techniques may well succeed in reconstructing the erasures more fully and more accurately. In looking in detail at each of the four maps in turn it is important to bear this in mind.[23]

Few traces of the first map of Palestine, on the recto page, are now visible **[26]**; we see most in the upper right quarter, least in the lower left. However, they are enough to show that it covered a smaller area than the successive maps of Palestine on the other side of the leaf and was a fairly simple map. While Jerusalem is drawn in the same form as on the other side, it is not identical: the hill there named Mount Sion is larger here, the tower of David smaller, and the base of each gate seems to protrude below the inner circle of the wall. Neither Mount Sion nor the tower of David is visibly named. The double circle of the wall was drawn with compasses: the hole made by the point is just above and to the right of the centre of the page. That the map has east at the top is clear from the few names of towns that can be read: Bethany above Jerusalem, Beersheba and Bethsura further to the right and, below them, probably Bethlehem and Hebron with what may be Ascalon below Jerusalem at the bottom of the map. Each name is written within an outline red symbol that can nowhere be seen clearly but that seems to resemble a building with a series of gable ends. There are traces of some nine more red symbols; on none can any letters be seen.

Above Bethany is a column of names without symbols which, from the few visible letters, may well be the same as those in the same position on the other side of the leaf: Gilgal (*Galgalis*), Jericho and the desert where the Lord fasted. Between Bethany and Jerusalem are the names of probably the mount of Olives – *mons* is visible and possible traces of a red symbol – and, though wholly illegible, the valley of Jehoshaphat, again as on the other side of the leaf. It seems that what we have is a map of the area between the Jordan at the top of the page and the Mediterranean at the foot, though all traces of both river and coastline have disappeared. This, however, ill accords with the map's sole clearly visible place-name that survives below the patch, *Caesarea Philippi*, which should be far from the coast, in the area of the source of the Jordan. The most likely explanation is that it was a mistake for Caesarea Palestina, which really is on the coast; confusion between the two did occur,[24] and in this case we can suppose the map's coastline making a sweeping curve to go round the hole in the parchment.

This was a map to be read as well as looked at, but their placing suggests that its blocks of text were added to the map where there was space, not written before the map was drawn. Of the four paragraphs on the left – three at the edge, one a little further in – no more than a few disconnected words can be read. A little more, however, can be seen of the three paragraphs on the right. They probably begin by saying that anyone who wishes to travel to Jerusalem should go to Ptolemaida – that is, Acre – and other discernible words later in the text include 'Golgotha', 'crucified'

23 I would here express my sincere thanks to members of the staff of the Department of Manuscripts at the British Library, who have given much help in my efforts to reconstruct the erased lines and words; I would especially mention the late Janet M. Backhouse, Dr Claire Breay and Mr Tony Parker. In 1997 the maps were viewed with video-spectral comparitor and with stereo zoom microscope, and their digitised images were enhanced in a variety of ways; nothing more appeared than was visible in good normal light and even ultra-violet light was slightly less effective. In 2006 they were viewed with Forth Photonics multi-spectral imaging system (MUSIS) which proved a little more successful.

Some errors on fig. 26 were detected too late to be corrected: 'I[e]r[ico]', '[ieiunauit]' and '[olivueti]' are the correct readings above Jerusalem, *é* at Bethlehem should be *e* with a line above, and there should be six, not five, lines of illegible text in the second block on the right.

24 Thus Antoninus of Piacenza, circa 570: 'Deinde venimus Caesaream Philippi, quae turris Stratonis, quae et Caesarea Palestinae vocatur' and variants (*Itinerarium Antonini Placentini: un viaggio in Terra Santa del 560-570 d.C.*, ed. by Celestina Milani, Università Cattolica del Sacra Cuore, Scienze Filologiche e Letteratura 7 (Milan, 1977), pp. 278-9). The passage occurs in the copy of this work that the mapmaker annotated (BL Add. MS 15219, f. 11r; p. 41 above); however, one of the short geographical texts that he himself copied in the same manuscript correctly locates Caesarea Philippi and clearly distinguishes it from Caesarea Palestina (BL Add. MS 15219, f. 2r).

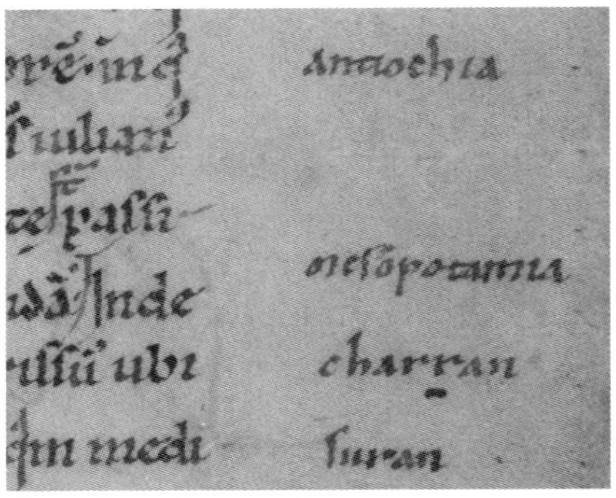

30. Marginal notes by the Tournai mapmaker in a copy of Antoninus of Piacenza's *Tour of the holy places* that belonged to his abbey, further evidence of his interest in the Holy Land. BL, Add. MS 15219, f. 11r.

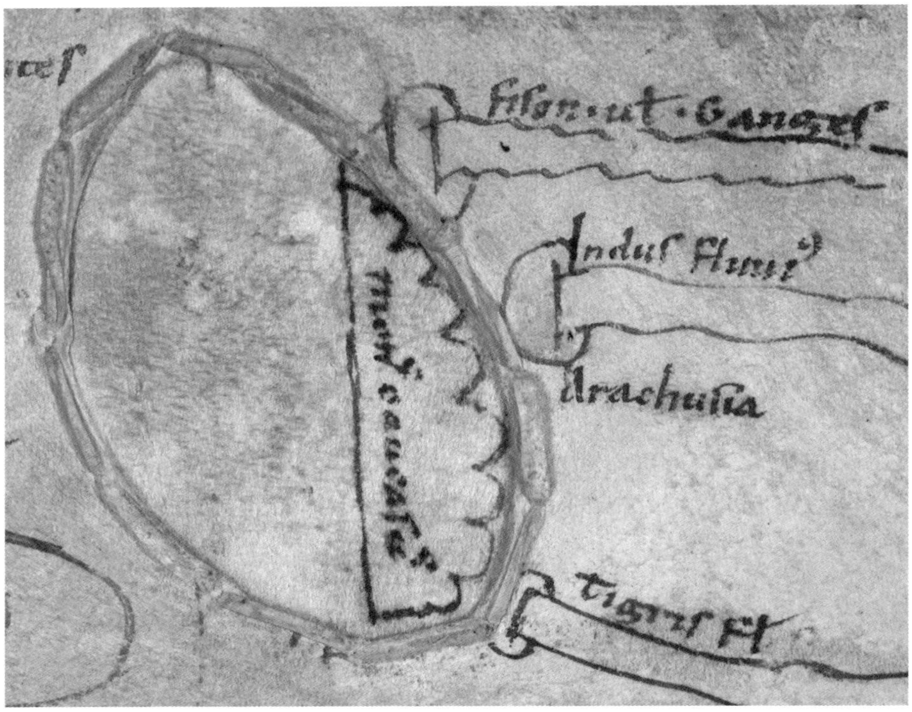

31. Parchment patch on the second map. The thin strip of parchment securing the patch minutely overlaps the sources of the Ganges and the Tigris, showing that the second map of Palestine was drawn before the patch was added for the map of Asia on the other side of the leaf. BL, Add. MS 10049, f. 64v.

and, probably, 'the church of the Cross outside the gate'. It seems to be a summary of the highlights from an account of the holy sites in Jerusalem, an account not unlike that which, as we have seen, our Tournai mapmaker copied in another manuscript, though it is not the same text. The rest of the map may well be just an extract from a world map with no further input; all the places that can be read, clearly or tentatively, can, for example, be found on the Ebstorf map except Bethsura and the desert where Jesus fasted. It is likely that the first Palestine map was quite a simple map.

The second map of Palestine, on the verso of the leaf, was more detailed and covered a much wider area, extending in the east to the Ganges and the land of the Scythians, and including most of Anatolia in the north, the Upper as well as the Lower Nile in the south **[28]**. It was clearly copied from a world map, very likely from the same world map as the first map of Palestine. The first map of Palestine, like the second, probably had a straight Mediterranean coastline and we have seen that its column of names above Jerusalem may very well be the same as on the second map. More certainly, the first and second maps of Palestine show Jerusalem exactly the same way, with Mount Sion and the tower of David beside the double circle with four gates – a form we know from no other map.[25] Equally telling is the place-name *Betsura*, Bethsura, one of the few names that can be read on the first map of Palestine; we see it also on the second,[26] but it appears on no other surviving world or regional map earlier than those based on the work of Burchard of Mount Sion from the end of the thirteenth century. We may suppose that the second map of Palestine was drawn as a supplement to the first, setting the area in its geographical context. It is certain that it was drawn before the map of Asia replaced the first map of Palestine of the other side of the leaf; the thin strip of parchment sewn around the patch attached to the recto, on the verso minutely overlaps the incomplete circles that mark the

25 Here there is no compass-hole at the centre of the double circle; presumably it was drawn around discs that had themselves been made with compasses.

26 It may, however, belong to the third, not the second, map of Palestine. It is written over an erased word, *P*[..]*g*[....]*l*; I take this to be a running correction on the second map, but it may possibly be the replacement of a name on the third.

sources of the Ganges and the Tigris, which must already have been drawn **[31]**.[27] The patch, we have seen, was attached when the first map of Palestine was erased and the map of Asia drawn in its place.

As we have seen, much of the second map of Palestine was erased, redrawn and rewritten to form the third map. It did not, however, suffer the total obliteration that overtook the first map on the other side of the leaf; some names and features of the second map continued untouched to form part of the third map. Reconstructing the second map thus meets two problems. One is discovering what has been erased. This is a little easier than in looking at the first map; the erasure has been less thorough. All the same, it poses considerable difficulty; many readings, especially at the north end of the map, are open to doubt and some words elude all attempts to read them. The second problem is to decide which of the lines and words on the third map are survivors from the second and which were added in the course of its alteration. Mostly it is fairly clear: *corozai*, near the mountains of Lebanon, presumably belongs to the third map, replacing the erased *choros[aim]* near Sebaste, now seen as wrongly placed. But it is always possible that this was a running correction: the name was entered first at one place then at another in the course of making the third map and did not appear at all on the second. Some running corrections we can in fact detect. Mount Gilboa has been twice erased, once (the name alone) just to the left of where it appears on the third map, once (with mountain symbol) very much closer to Jerusalem **[32]**. It was the latter that belonged to the second map; its name close to its present position was entered for the third map, then moved to make way for the river Kishon. But exactly what happened is not always as clear as this and we must bear in mind that a certain amount of guesswork – at best, informed guesswork – has gone into the reconstruction of the second map.

The second map gives Palestine a completely straight coast from St Symeon at the mouth of the Orontes in the north to Alexandria in the south, where it turns west towards the mouth of the Nile. There the new Babylon lies a little way inland and a mountain range, *mons Catapathimon*, for some the boundary between Asia and Africa, extends to the Upper Nile;[28] beyond it a deleted (and illegible) word seems to be the one point on the whole leaf where erasure went too far, making a tiny hole in

27 This, however, implies that the course of the Tigris was redrawn earlier than the changes that created the third map of Palestine; changes on the verso of the leaf may well have been more piecemeal, less systematic, than I suggest here. I have considered the possibility that the parchment strip was flattened in the course of conservation in modern times, and that only then did it overlap the sources of the Ganges and the Tigris, but on careful examination I do not see that this could have happened.

28 *Mons catapathimon* is discussed by Kugler, *Ebstorfer Weltkarte*, I, p. 184.

32. Mount Gilboa on the second and third maps. Mount Gilboa, just left of Jerusalem on the second map, was erased in making the third map and its name (without mountain symbol) written rather further north; at once it was again erased and entered (with symbol) nearby to make room for the river Kishon. BL, Add. MS 10049, f. 64v.

the parchment. East of St Symeon the southward then eastward bends of the Orontes give room to show Anatolia and neighbouring regions in some detail, with boundaries drawn in and provinces named. No part of the Black Sea appears unless in the uncertain reading [*ci*]*meric*[*um m*]*are*, partly cropped at the edge of the page. Further east are Hircania and the Caspian Sea. The Ganges and the Indus flow south and between them is India, with the columns of Hercules and Alexander and the oracle of the sun and the moon. The Indus and, further west, the Tigris and Euphrates all flow separately into the Red Sea, where there is a large island, the isle of the Gorgons. Palestine is clearly demarcated, bounded on the north by the mountains of Lebanon, which close it off from the Orontes, on the east by the Jordan and on the south by the Lower Nile. The sea of Galilee lies on the Dan before its union with the Jor, and beyond the Dead Sea the Jordan flows into the Red Sea, as on the Ebstorf world map. In Palestine towns along the coast are named in an unbroken series, but inland besides Jerusalem there is what can only be described as a scattering of other towns. They include all those that can be identified on the first map of Palestine, but apart from Bethsura this is not significant; Bethany, Beersheba, Bethlehem and Hebron would be likely to appear on any map. Some mountains are named, with humped symbols, but there are no rivers between the Jordan and the Mediterranean.

It is not only at Jerusalem that we see a pictorial element in the second map of Palestine. Four stylised trees mark the forest of Hircania (*ircania silua*) and two, of a different kind, the oracle of the sun and the moon, while the columns of Hercules and of Alexander are drawn with three columns each. However, Noah's ark is not pictured, though there is a note that it is on the mountain of Armenia (*mons Armenie*). Apart from Jerusalem the only town marked with a symbol is St Symeon, where there is a battlemented rectangle; even in the series of towns along the coast there are no frame-lines to separate one name from the next. Where a river does not have its source in a mountain it starts with an incomplete circle to represent a spring – the same convention appears on some other maps, the Munich and Psalter world maps among them – but those of the Jor and Dan have been erased as a running correction to make way for the mountains of Lebanon. The other springs have been painted yellow, and so too have the Red Sea, the bases of the columns of Hercules and Alexander, the mountains of Lebanon, the mountain of Armenia and the *mons Catapathimon*.

The map of Asia on the recto is vastly more complex and sophisticated than the first map of Palestine that it replaced. It covers a huge area, from the Peloponnese and the Balkans at the foot of the page to the Ganges, Sri Lanka (*Taprotane*) and the Indian Ocean at the top. On the south it extends into Palestine as far as Jaffa and the union of Jor and Dan to form the Jordan, and the tip of the Persian Gulf protrudes, painted red. The northern – left-hand – edge of the map is slightly damaged, but we see the Caspian Sea and the whole of the river Don (*Tanais*). Dominating the lower half of the map, and covering about one-fifth of the entire page, is Anatolia, divided into its provinces.[29] Regions and provinces are named elsewhere on the map – *Galilea*, *Parthia* and others – but only here are the lines of their boundaries drawn in where they are not marked by rivers. Here and throughout the map, where a river does not

29 A full explanation and discussion of the map's representation of Anatolia, and its wider implications, is given by Stéphane Lebreton, 'Cartographier l'Asie Mineure dans l'Antiquité: pour quelle utilité? La carte 1 du manuscrit de saint Jérôme', *Res Antiquae*, 2 (2005), pp. 265-94.

originate in a mountain its source is shown, as on the second Palestine map, by an incomplete circle that must represent a spring. Mountain ranges appear as lines of humps, single mountains as triple humps; the flame-like points on the Caucasus, the Caspian Gate and *Sephar mons* may signify particularly high peaks.[30]

The way towns are shown deserves much further investigation. Many are simply named without a symbol, especially in the lower half of the map, like *Thebe c* in Greece and *Corozaim* in Palestine. Many others are shown with a symbol resembling an open circular structure, with cornice around the top, viewed from an oblique angle. In the upper half of the map, and also at Constantinople, these structures have doors and windows; on some these are coloured red, and on some so too is the cornice. It looks rather as if the mapmaker began to elaborate all these symbols in this way, starting at the top, but failed to extend the operation to the lower half of the map except at Constantinople, too important to be left unadorned. Indeed, Constantinople has also been given three spires, while Babylon has a tower beside it. However, nine towns in the upper half of the map are marked with a completely different symbol, a rectangle battlemented at the top; a tenth lacks the battlements, perhaps because space did not allow. All but two of these rectangles are drawn in red. Here again the mapmaker may have meant to add this symbol to the names of towns in the lower half of the map, though space would have been an even greater problem here.

Besides physical features and towns the map presents a little further information. Noah's ark appears in Armenia, and north of the Black Sea three squares mark the three altars of Alexander, set at the northern limit of his conquests; the three at the eastern limit, which might well also be shown, do not appear. By the mouth of the Danube we are told of the pygmies who fight against the cranes, and we have seen that on Crete, formed by the parchment patch now threaded on to the page, we have a note of its hundred towns and of its status as a Greek province. There are notes too on islands in the Indian Ocean: Sri Lanka is divided down the middle between the half that is habitable and the half that is not, we are told that gold abounds on *Crise*, silver on *Argire*, while on an unnamed island live the *Hipode*, a people whose legs are those of horses. Finally, along the top of the page, three winds of the eastern quarter are named in cartouches.[31]

This confirms what we would anyway have supposed: the map is an extract from a map of the world, but this was not the world map from which the second, perhaps also the first, map of Palestine was copied. As we shall see when we look at the final state of the map of Palestine on the other side of the leaf, the mapmaker probably had the whole world map before him when he copied this portion; he did not copy an existing map of this portion alone. But further evidence, surprising and interesting, has recently come to light. Patrick Gautier Dalché has identified in the Biblioteca Nazionale at Naples a twelfth-century manuscript describing a world map that corresponds to the map of Asia in many details, too many to be due to chance; the map described is the map from which the Tournai mapmaker copied his extract. It has, for instance, the same note on Crete, the same notes on the islands of *Crise* and *Argire*, the same division of Sri Lanka into two zones, one inhabited, with ten cities, the other uninhabited because of its dragons and elephants. The correspondence is not complete, and indeed we could hardly expect it to be: the writer of the text need

30 *Sephar mons* is discussed by Westrem, *Hereford map*, p. 64; both on the Hereford map and here, it has a namesake at the source of the Indus.

31 The three relevant winds, Vulturnus, Subsolanus and Eurus are discussed (with reference to their appearance on the Hereford world map) by Westrem, *Hereford map*, pp. 12-15.

not have described every feature of the map before him, nor need the mapmaker at Tournai have copied every detail. In Greece, for instance, both map and text name Athens and Thebes, but Corinth appears only on the map, while by Constantinople the text names the Hellespont but the map does not. However, there is enough coincidence between map and text for us to be certain of a connection between the two.[32]

This is not to say, though, that mapmaker and author both saw the same actual map. In a careful and detailed argument Gautier Dalché suggests that the Naples manuscript is very likely a copy of a text written not later than the eleventh century at the abbey of Bobbio in Piedmont. What it describes is a copy of a world map otherwise known to us only from a 1461 catalogue of the library at Bobbio. This map was 'in the Irish style' (*secundum scoticam traditionem*), and it included, surprisingly, what Gautier Dalché has identified as Iona and other isles in the Hebrides. He raises the possibility that the map, drawn between the sixth century and the eighth, came to Bobbio from Iona; one need not look far for Bobbio's links with Ireland and the Hebrides, for Columban, who founded the abbey in 613, himself came from Ireland. However, the text describing the map also gives a particularly detailed account of Egypt; Gautier Dalché suggests that a map from a continental source was brought to Iona, whence a copy with local detail added found its way to Bobbio, and tentatively he points to Eucher, bishop of Lyon, who died in 450, as the possible author of the original map. What our mapmaker copied at Tournai in the twelfth century could well have been a world map of insular origin, a more or less exact version of the map that was at Bobbio; there would be no great difficulty in explaining how it might have reached Tournai. Equally, though, it may have been a version of the original map without the added insular names; there is no way of telling which. We can, however, be sure that the map described in the Naples manuscript was closely related to the map from which the Tournai mapmaker copied his map of Asia.

We have seen that some alterations were made to the map of Asia. The names of nine towns or regions have been erased and rewritten in a new position: *humscite* in the north-east, *hibe*[*ri*]*a c* and *Ar*[*men*]*ia* further south, *Comagena* and *Siria sodal* west of the Euphrates, *tiatira c*, *derbe ciu*, *atalia c'* and *listra c* in Anatolia. In the Indian Ocean a lozenge-shaped island east of the mouth of the Ganges has been erased and changes have been made to the estuaries of two rivers, the *Octorogorra* in the north and the *Hipanis* in the south – in the latter case at least twice. A short range of mountains has been erased west of the source of the Tigris and Euphrates, and in Greece the coastline north of Athens has been slightly altered. At the very foot of the map what was possibly *Triesta c* has been erased and not replaced, perhaps because it was found to lie beyond the edge of the page, and at the top *s oriens*, presumably for *sol oriens*, 'rising sun', has been erased and perhaps replaced by the three names of easterly winds. All these changes can be seen as running corrections made in the course of copying the map; none need be a later alteration.

The radical changes that converted the second map of Palestine into the third map introduced no further colour to the verso page – all were made in black ink. The reason for these changes was clearly to reconcile the map with the map of Asia on the recto, so that it did not give a contradictory picture of the areas common to both, and

32 Patrick Gautier Dalché, 'Eucher de Lyon, Iona, Bobbio: le destin d'une *mappa mundi* de l'antiquité tardive', *Viator*, 41 (2010), pp. 1-22. I am most grateful to Prof. Gautier Dalché for sharing this discovery with me before its publication.

at the same time to bring the rest of the map into line with the world map from which the map of Asia had been copied. The two maps, one on each side of the leaf, would then present a substantial portion of this world map, a version of the world map described in the Naples manuscript. Relatively few changes were made in the upper third of the map, east from the Euphrates; there was no attempt to introduce the mass of detail that we see here on the map of Asia. The few details there already and allowed to remain may perhaps have been on the world map, though not copied on to the map of Asia; the pillars of Hercules and of Alexander are both named in the Naples manuscript. The map's focus of interest was still Palestine with the neighbouring lands to north and south.

The most obvious change was to the straight Mediterranean coastline. A deep gulf to the north now separates Palestine from Anatolia and another, rather smaller, named the Egyptian Sea (*egiptium mare*), from Egypt in the south. The Lower Nile has lost two semi-circular islands but at its reconfigured mouth has gained a large island that corresponds to its delta. Here, and further east along the Lower Nile, names have been added reflecting the interest in Egypt that appears in the Naples manuscript. Alexandria, on a new site, is now marked with a triple-spired structure like Constantinople on the map of Asia, with beside it a tower identified as the 'very tall lighthouse' (*farus altissima*);[33] to the north is Abukir, shown as an island (*Canopos insula*).[34] The city and river of *Rinocorura* – El Arish and Wadi El Arish – are now shown, with a note that the river marks the boundary between Palestine and Egypt; the town and, further east, the drastically re-sited new Babylon are both marked with the map of Asia's symbol of a circular structure, plus a central tower. Along the now curved Palestine coast a new series of towns is named, not exactly the same as those they replace on the second map's straight coastline; Yibna (*Iamnia*) is a new name, for instance. Each is now framed in a rectangle with a protruding square tower on top, a simplified form of the battlemented symbol used for nine towns on the map of Asia. At the northern end of the coast the names are the same as on the map of Asia, except that Haphraim (*Capharnaum*) and Caesarea are interposed between Acre and Jaffa.[35] St Symeon and the Orontes have disappeared, but in the gulf between Palestine and Anatolia and further west are the same islands as on the map of Asia, some with island outline drawn around the name, some without, and differently arranged – as we shall see on other maps, what mattered was to place names in the right area rather than in the right relationship one to another. Anatolia is now bounded on the north by the Black Sea, on the west by the Dardanelles, with Constantinople marked by the same symbol as on the map of Asia (though with towers instead of spires).

From Anatolia all names and boundaries have been erased, but no attempt has been made to introduce the pattern of provinces that appears on the map of Asia – wisely so, as the result could well have been extremely messy. The Taurus Mountain (*mons Taurus*), however, remains intact. To the south the Red Sea has been truncated to the west but extended to the north with the addition of the Persian and Arabian Gulfs (*Persicus sinus, Arabicus sinus*); the yellow has been fairly effectively removed from the lopped-off western end, but the added gulfs have not been coloured. The course of the Euphrates has been changed so that it now shares a mouth with the Tigris on the Persian Gulf, and the Dead Sea now closes the Jordan; it no longer

33 In the Naples manuscript it is *pharos turris altissima*.

34 As on the Hereford world map and elsewhere (Miller, *Mappaemundi*, III, p. 18; Westrem, *Hereford map*, pp. 400-1).

35 Miller, *Mappaemundi*, III, p. 14, convincingly identifies *Capharnaum* (his reading of the word is not quite correct) as Haphraim; at some point in the map's transmission a familiar name was substituted for an unfamiliar one. The real Capernaum on the sea of Galilee also appears on the map (as *Cafarnaum*), a survivor from the second map of Palestine. The same confusion probably lies behind the location of *Cafarnaum* a little way east of Caesarea Palestina on the Oxford map of Matthew Paris. Haphraim and Caesarea may well have been omitted from the map of Asia to give room for Jaffa to be included.

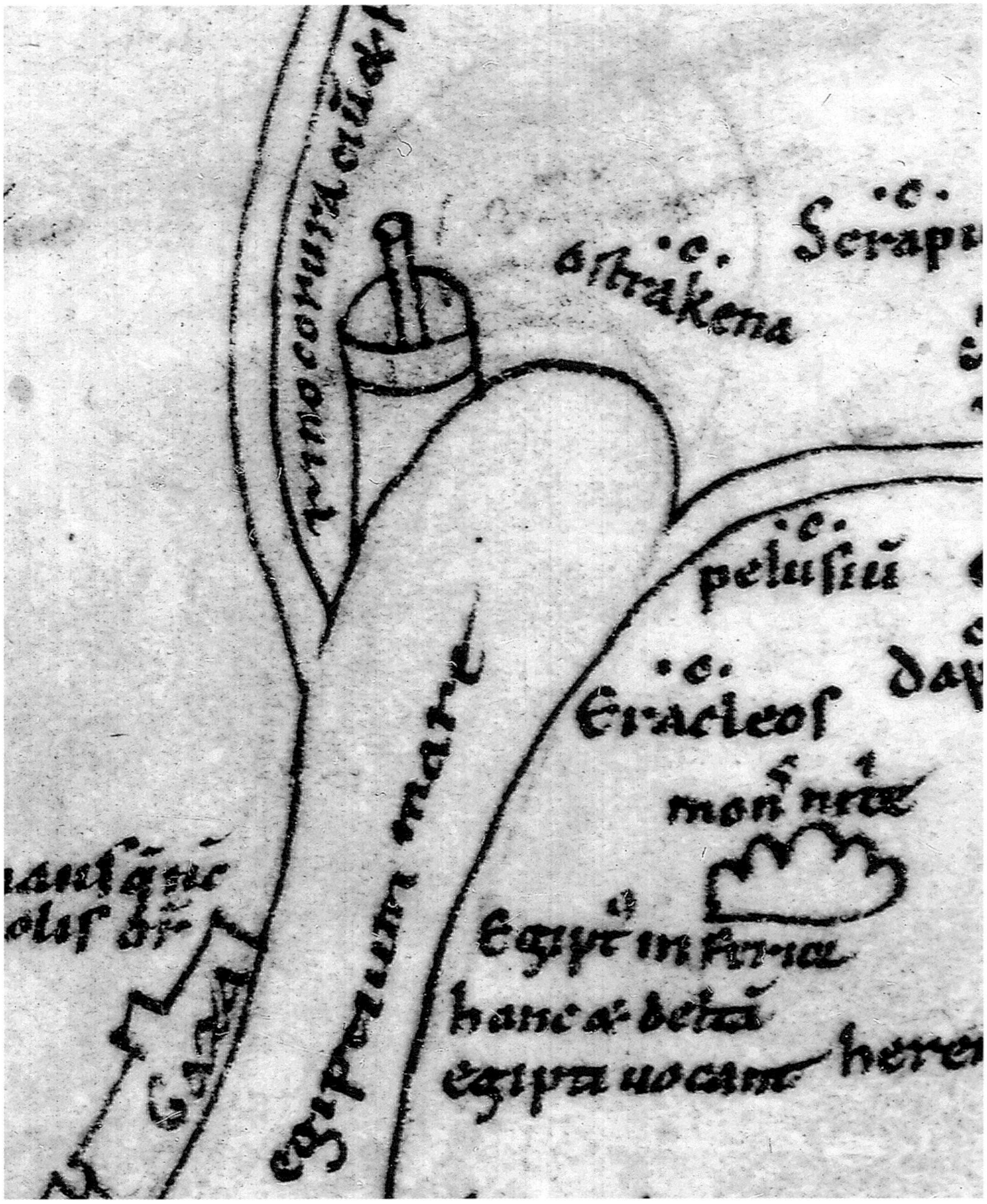

33. Drafting lines by the Egyptian Sea on the third map. As first drawn for the third map, the gulf south of Palestine named the Egyptian Sea extended further east and erased lines, just visible, show a series of changes made before it reached its present shape. BL, Add. MS 10049, f. 64v.

flows into the Red Sea. At the other end of the Jordan the Jor has been redrawn on a new course, and a new river, the Kishon, has been added to flow into the sea of Galilee. Throughout the northern part of Palestine names of towns and mountains have been erased and mostly entered again in new locations; the relative positions of places seem to have mattered more here than in the Aegean. The replaced mountains are marked with hump-like symbols, but no symbol marks any of these inland towns – they are simply named.

Overall one gets the impression that the conversion of the second map of Palestine into the third map was carried out with perhaps a little more haste, a little less care, than went into the map of Asia or, indeed, the second map itself. We have noticed the small running correction made at Mount Gilboa beside the river Kishon. In two other places the mapmaker had trouble in drawing the new lines for the third map. One was in the Egyptian Sea, which extended further east when first drawn and was then twice shortened before the river and town symbol of El Arish were added **[33]**.

The other was at the south end of the Dead Sea, where a confusion of erased lines defies reconstruction of the sequence of drafting **[34]**. And here again it must be stressed that in many cases what belonged to the second map of Palestine, what was erased or added for the third, can only be a matter of judgement or, indeed, of simple guesswork. However, the contrast between the careful draftsmanship of the map of Asia and the scratchy emendations that created the third map of Palestine makes it unlikely that they belong to a single operation. Perhaps, after some time, the mapmaker's attention was drawn to the discrepancies between the map of Asia on the recto and the second map of Palestine on the verso, and it was only then that he made the changes that turned the second map of Palestine into the third.

The third map of Palestine did not quite complete the Tournai maps as we know them. Eight place-names have been added, five on the recto and three on the verso, in several different hands and over a long period. On the map of Asia *ceneris porta corintiorum* was probably added not long after the map was drawn but *Troade c* and *beroca c*, as well as *bethel*, *Silo* and *lebona* on the map of Palestine, may well date from the fourteenth century, and *pataram* and *neapolis c* on the map of Asia even from the fifteenth. None of these names occurs in the Naples manuscript but this need not be significant; they might still have appeared on our mapmaker's exemplar – or on some other map. It is extremely interesting that the maps were not only consulted and closely examined but were compared with other maps or texts long after they were drawn, for this tells us something of the growing awareness of cartography in the later Middle Ages.

All that is offered here is a first attempt at unravelling the tangled composition of the Tournai maps. That they are not the work of Jerome makes them no less interesting and important. Thanks to the Naples manuscript we can be sure that, properly interpreted, they set before us a considerable portion of a world map of great antiquity. And thanks to the careful economy with which the monks of St Martin's abbey husbanded their supplies of parchment eight hundred years ago, ensuring that elaborate erasure was preferred to using a new sheet, we are given a glimpse not just of one early world map but of two.

34. Drafting lines by the Dead Sea on the third map. The shape of the Dead Sea seems to have been changed several times in turning the second map of Palestine into the third, but the erased lines are confused and some of the lines may come from running corrections when the second map was drawn. BL, Add. MS 10049, f. 64v.

7 THE OXFORD MAP OF MATTHEW PARIS
mid-13th century

Oxford, Corpus Christi College, MS 2* (formerly MS 2), ff. 1v, 2r **[35]**. Parchment, 34.5 × 48 cm. Dark brown and red ink.[1]

THE NAME AND WORK of Matthew Paris are well known to every historian of thirteenth-century England: his chronicles are a crucially important source for events in England in the thirty years before his death in 1259.[2] He does not, however, tell us a great deal about his own life. It was spent as a Benedictine monk at St Albans Abbey, twenty miles north-west of London, which he entered in 1217; this means he was born probably in about 1200. His name need have no connection with the city of Paris, but it may just possibly mean that he had studied there.[3] His only recorded travel outside England was to Norway in 1248-9, when he was sent, with papal authority, to reform the Benedictine house at Niderholm, which had got into financial difficulties. The mission is the more surprising in that he apparently never held any administrative office in the abbey at St Albans.

He had, however, an extraordinary range of acquaintance among important people in England. This was almost ensured by the position of St Albans at a junction of routes near London and Westminster: the abbey would receive visitors on their journeys to and fro. King Henry III stayed there on nine occasions between 1220 and 1259 and he was only one of many who met Matthew, talked to him and supplied him with information and copies of documents for the history that they knew he was writing. It must have been through these contacts that Matthew came to be present at some of the ceremonial events he describes, among them the presentation to Westminster Abbey of a vessel containing the blood of Christ in 1247 and the marriage of Henry III's daughter to King Alexander III of Scotland at York in 1251. He was not, however, a sycophantic chronicler of the doings of the great, and his own views, his own personality, constantly intrude on his narrative. Thus his acquaintance with Henry III, with whom he talked at length, did not stop him from criticising the king in writing about him, and when he came to prepare an abridged version of his principal chronicle for circulation outside St Albans, he marked many passages – as we still can see – as unsuitable for a wider public.

But Matthew was not only an historian and commentator; he was much else besides. Combined with a lively, enquiring mind he had considerable graphic skills. He was indeed an artist and his tinted line drawings, of which the best known is of the Virgin and Child with himself kneeling at her feet, have a significant place in thirteenth-century English art.[4] It is typical of him that, having seen the elephant that King Louis IX of France gave to Henry III in 1255, he both described it and drew a picture of it, and there are many illustrations in his chronicles. These range from

1 I am most grateful to Mrs Christine Butler and Dr Julian Reid, successive Archivists of Corpus Christi College, Oxford, for their very positive help when I have examined the manuscript; also to Prof. Evelyn Edson for giving me a copy of her unpublished catalogue of the places named on the map, which has been most helpful.

2 Accounts of Matthew's life and work are in Vaughan, *Matthew Paris*; Gransden, *Historical writing*, pp. 356-79; and, by Simon Lloyd and Rebecca Reader, in the *ODNB*, XLII, pp. 620-8.

3 While more often deriving from the place-name, Paris is also a patronymic from a form of Patrick (P.H. Reaney, *A dictionary of English surnames*, ed. by R.M. Wilson (3rd edn, London, 1991), p. 338.

4 However, it is no longer thought that he headed a school, a group of illuminators, at St Albans as suggested by M.R. James, *La estoire de seint Aedward le rei*, Roxburghe Club (Oxford, 1920), pp. 17-34; later work on his art in general includes Vaughan, *Matthew Paris*, pp. 205-34; George Henderson, 'Studies in English manuscript illumination, part I', *Journal of the Warburg and Courtauld Institutes*, 30 (1967), pp. 70-85; Nigel Morgan, *Early Gothic manuscripts*, Survey of Manuscripts Illuminated in the British Isles 4 (2 vols, London, 1982-8), II, p. 22; Lewis, *Art of Matthew Paris*.

elaborate drawings of battles and sieges to simple symbols in the margin: for a coronation, for instance, there would be a crown and shield, for the death of a noble a drawing of his shield upside down. He was used to expressing himself visually, in graphics of every kind, and his manuscripts include not only pictures but diagrams, of the winds, of Roman roads in England, for fortune-telling and so on. It is in this context that we should see Matthew's maps, the works of an imaginative man of wide interests and a natural draftsman.[5]

As we shall see in looking at his Acre maps of Palestine, most of Matthew's maps are in the manuscripts of his chronicles. However, the map of Palestine that belongs to Corpus Christi College, Oxford, is an exception. It is on a bifolium that was bound at the beginning of a thirteenth-century Bible but is now kept separately; the book belonged to St Albans and is partly in Matthew's hand. Names in the manuscript show that before it reached Corpus Christi College it had belonged to Morgan Kidwelly, Speaker of the House of Commons, who died in 1513, and to John Tyson, rector of St Nicholas's church at Worcester from 1532 to 1544.[6] The bifolium was prefixed to the volume not for the sake of the map but for the two unfinished pictures on the other side, showing Christ's Deposition from the Cross and the three Marys at the sepulchre; indeed, as bound in the book the map could not be seen as a whole, for it occupied the recto of the first leaf, the verso of the second. These two pictures, probably intended for a psalter, have been described as masterpieces; they date from about the 1140s and have been ascribed to St Albans on the basis of Matthew's map on the other side.[7] As we shall see, this is not necessarily a safe argument, though the ascription may well be correct.

In the 1240s the pictures' quality was apparently not recognised, and the back of the parchment was put to less elevated use **[36]**. First, the bifolium was folded a second time, dividing the sheet into four, and on one of the outer quarters two short texts were written, each in a different hand, neither of them Matthew's.[8] The first is a document sent to the pope in March 1246 from a meeting of prelates and lay magnates in London, together with accompanying letters from the bishops of the province of Canterbury, from the abbots and priors and from the earls, barons and all the people of England. The document is a list of grievances, seven paragraphs, each starting 'It is a vexation . . .', 'Again it is a vexation . . .' (*Gravatur, Item gravatur*). It is known from other sources, and Matthew himself includes the text in his principal chronicle, the *Chronica majora*, but here it is a slightly different version that may be an early draft of the text that was actually sent. The second text, otherwise unknown, describes some of the discussions that led to sending these letters.[9]

These texts contribute to our understanding of Matthew's map. He drew it on the other three-quarters of the sheet, giving it a curious L-shape: the whole of the page on the left, the lower half only of the page on the right. Clearly the texts were there already when Matthew drew the map, so he must have drawn it later than March 1246. It follows too that he did not view the map as a fair copy, a finished product, worthy to accompany the pictures on the other side in prefacing a psalter or other fine volume. Rather, he seized on the first piece of discarded parchment he could get of the right size (if not the right shape) to make a rough copy of an existing map of Palestine. The map is not Matthew's own creation. There are many indications that he copied

5 Lewis, *Art of Matthew Paris*, includes (pp. 321-76) a chapter on his maps, but covers only those that accompany his principal chronicle, the *Chronica majora*; Connolly, *Maps of Matthew Paris*, is concerned particularly with his strip maps of the route from London to south Italy, and looks only briefly (pp. 150-2) at the Oxford map of Palestine.

6 Information from the College's unpublished catalogue of its manuscripts.

7 T.S.R. Boase, *English art, 1100-1216* (Oxford, 1953), p. 178; C.M Kauffmann, *Romanesque manuscripts, 1066-1190*, Survey of Manuscripts Illuminated in the British Isles 3 (London, 1975), pp. 101-2; Rodney M. Thomson, *Manuscripts from St Albans Abbey 1066-1235* (2 vols, Woodbridge, 1982), I, pp. 31, 112; II, plates 117, 118; Alexander and Temple, *Illuminated manuscripts*, p. 6.

8 The clearest of many details is that the forms of the ampersands in both texts differ from those written by Matthew, shown in Vaughan, 'Handwriting of Matthew Paris', plates XV-XIX.

9 Both texts are published, the list with variants from other copies, in *Councils and synods with other documents relating to the English Church*, II: *A.D. 1205-1313*, ed. by F.M. Powicke and C.R. Cheney (Oxford, 1964), part I, pp. 395-8. I cannot agree with these editors that the documents are in Matthew's hand (pp. 395, 397n).

Anthiochia
Baruth
Nazareth
Cesarea

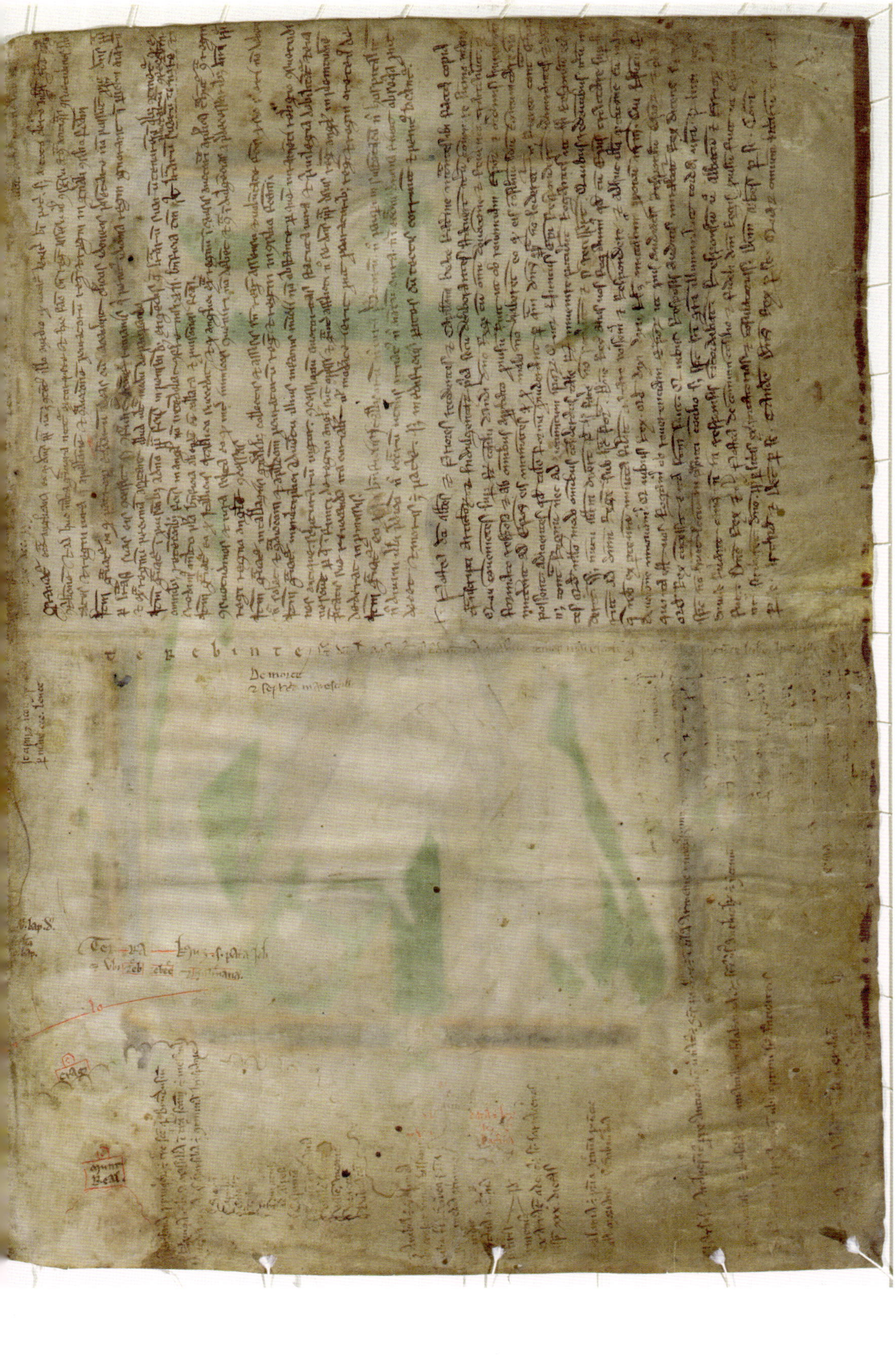

35. The Oxford map of Matthew Paris, mid-13th century; north at the top. The map occupies the whole of the left-hand page, the lower half only of the page on the right. The thick dark lines and other coloured areas on both pages come from the pictures on the other side of the leaves. Oxford, Corpus Christi College, MS 2*, ff. 1v, 2r.

it, with increasing haste, from a map to which he had only limited access, and there are some indications that he saw it as a sketch or cartoon from which he could work up a finished copy of his exemplar.

At first sight the map is entirely dominated by a thick brown rectangle, with twelve gaps in it, that outlines the central portion of the left-hand page where we see most of the map's detail. This rectangle has nothing to do with the map, and we must try if we can – and it is not easy – to see the map without it. It is in fact the frame of the picture of the three Marys on the other side of the parchment, broken by six roundels containing anonymous heads and six foliate ornaments. On the right-hand page the

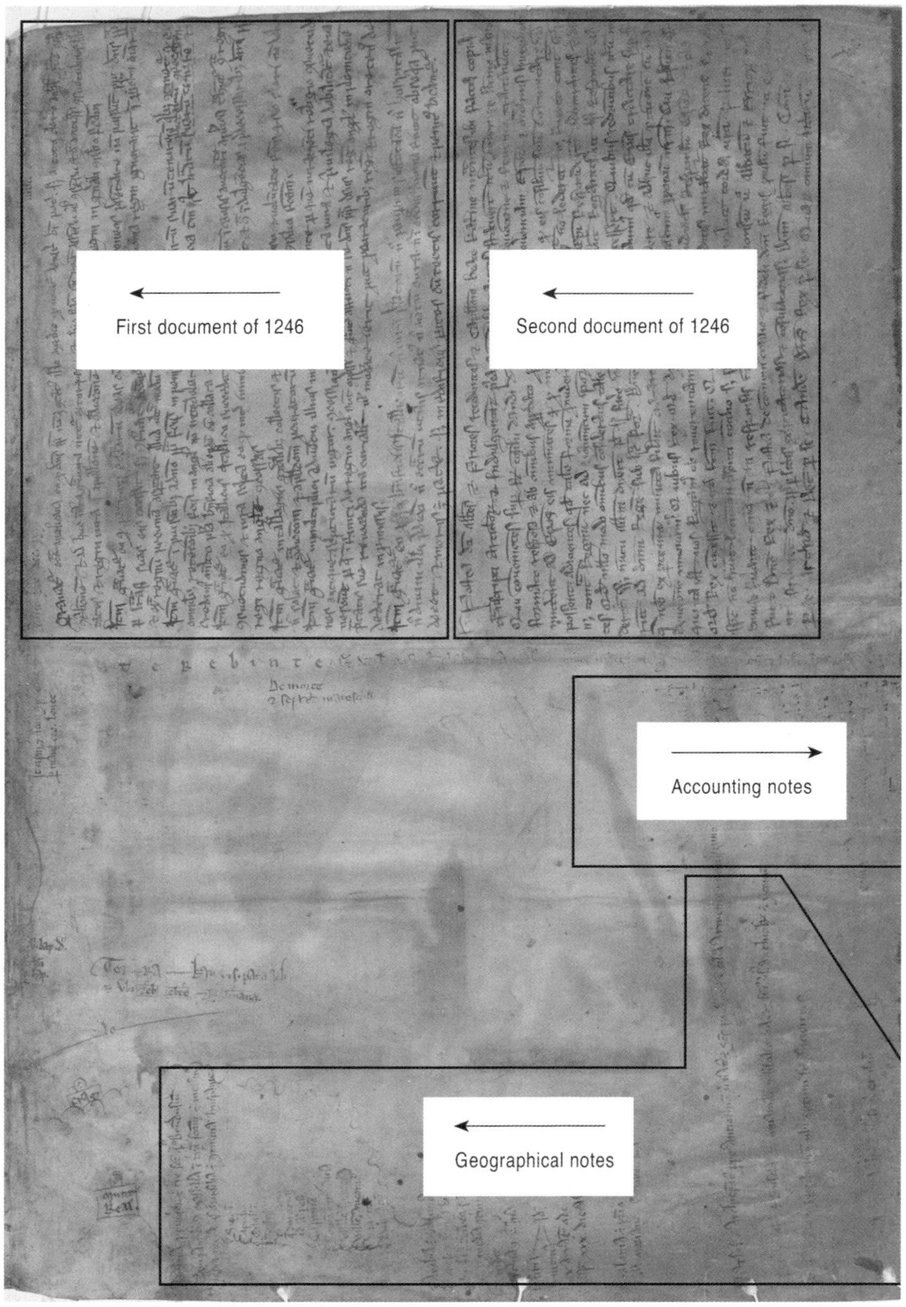

36. Plan of the right-hand page, showing the positions of the two documents of 1246, the geographical notes and the accounting notes.

similar frame of the Deposition does not show so clearly through the parchment but on both pages we see without difficulty the figures in the pictures on the other side. It is likely that the frame on the left-hand page could already be seen through the parchment when Matthew drew the map, but not so overwhelmingly as now; it has probably become more emphatic with the passage of time. Thus Mount Carmel and its surrounds seem almost tailored to fit the frame's central gap on the left side, and likewise the figure of Lot's wife turned to salt in the lower right corner. On the other hand many details, among them the names of the towns along the sea of Galilee, are now difficult to read because they are written on or across the line of the frame. It cannot have been so startlingly evident when Matthew drew the map.

North is at the top of the map and the Mediterranean coast follows a more or less straight line along its left edge, from St Symeon at the mouth of the (unnamed) Orontes down to Alexandria and Cairo in the south. The map's layout gave Matthew some trouble. He first drew the coastline rather more than a centimetre further to the right, then, perhaps realising he was going to need as much space as possible, he erased it and redrew it closer to the edge of the parchment. There are clear signs of this erased line in the north between St Symeon and Sidon, and just above the name of Damietta in the south we can see how this name was first written further to the right and then erased. Matthew must have entered Jerusalem on his copy at an early stage, with other places around it, but without its local holy sites. When these were added, Emmaus had to be erased and placed further west, and the position of Gaza, originally south of Emmaus, may have been changed at the same time **[37]**. Further north and east there were greater difficulties. Damascus also had been entered early on, together with the rivers Abana and Pharpar and a note that they flowed simply into the desert sands. When in due course Matthew came to enter the Jor and

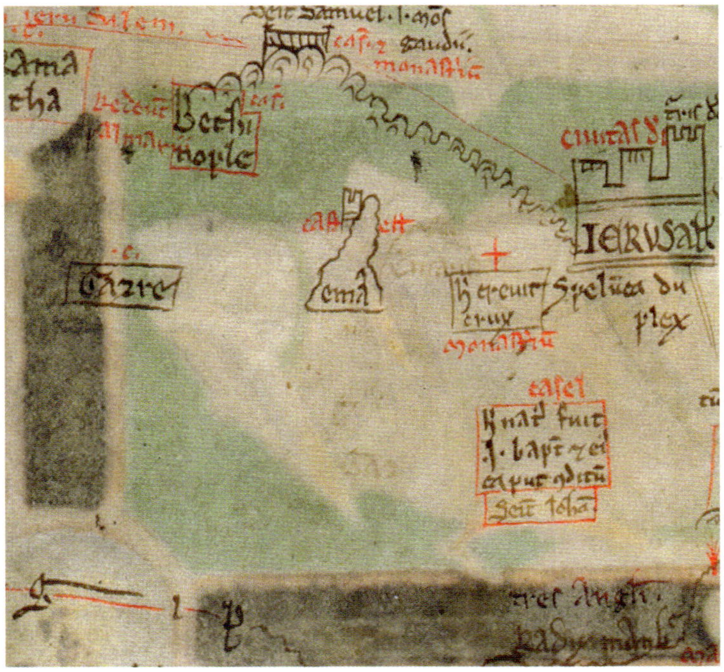

37. Corrections made in drawing the map. Emmaus and Gaza were first entered rather closer to Jerusalem, then moved further west so that Emmaus did not impinge on the other sites now entered around the city.

the Dan, the two streams that joined to form the Jordan, he found that Damascus should have been placed much further to the east – further, indeed, than the parchment would allow, for it was here covered by the texts of 1246. All he could do was to make the Jor and the Dan also flow through Damascus, with a note that they did not really do this and should have been drawn nearer the coast (*Istud propinquius est mari nec contingit damascum*). The miscalculation over Damascus affected the rest of the Jordan: both the sea of Galilee and the Dead Sea are drawn much too far to the south, as they had to avoid the note about the Abana and the Pharpar. The Jordan valley is thus quite out of relation to the rest of the map: either the changes needed were too great to be dealt with by erasure and redrawing or, simply, time was too short.

It is when we look more closely at its detail that we see how the map was copied with increasing haste. This is perhaps why Arqa in the north-west was first entered on the coast; still in the course of copying, it was erased and placed, correctly, inland. There is a little cluster of holy sites around Jerusalem, but, broadly speaking, the further we move from north to south, the further from west to east, the less detail appears and the map becomes more and more sketchy. It is only along the coast that we are given distances between places, and there only as far south as Caesarea Palestina. Apart from a note that mentions Lydda as a bishopric, bishops' sees are identified only along the coast from Sidon northwards. Again, though some religious houses are noted elsewhere it is only Belmont in the north-west that is assigned to a particular order, the Cistercians, and the three castles whose owners are named are all in the north-west: Chastel Blanc (struck out), Montferrand and Krak des Chevaliers. Matthew had red ink as well as black to hand throughout the work of copying – the red was not entered as a separate operation – and he started at the top with systematic use of the two colours, but this was apt to lapse in the map's farther reaches. Thus black, not red, is used to frame the names of the towns along the sea of Galilee, Gaza and the site where the wood of the Cross grew. In the top half of the map regional names are written in black letters spaced out with a broken red line between each letter or syllable; below, 'The land of Egypt' (*Terra Egipti*) is struck through with a continuous red line followed by a broken black one, while 'The land of the sultan of Babylon' (*Terra soldani Babilo*), all in red, lacks the final *nie* – just as, higher up, the final *e* is missing from 'The land of Syria' (*Terra Siri*). These are all tiny points, but taken together they give the impression that Matthew was working faster, less carefully, as he proceeded down the map.

Some of the map's mistakes may point the same way, but others were almost certainly already there on Matthew's exemplar – he simply copied what was before him. Thus it is unlikely that it was Matthew who misplaced Margat south of Tripoli instead of north of Tortosa. On the other hand he may have been responsible for reversing the names of Montferrand and Krak des Chevaliers, and of Nablus and Sebaste; whether they originated on Matthew's copy or were on his exemplar, these two mistakes must have arisen from confusing the names of two sites placed close together. Certainly it was Matthew who made the slip of the pen in describing the river that divides Syria from Palestine: he wrote Palestine for Syria, again a sign of haste.[10] It was probably Matthew who gave the name Tigris to the Nile at Damietta;

10 'This river, which is small, divides Syria from Palestine, that is, the Holy Land, which is to the south, and Palestine, which is to the north' (*Iste torrens qui paruus est diuidit Siriam a palestina id est terram sanctam que est uersus austrum et palestinam que est uersus aquilonem*).

Evelyn Edson has pointed out that this branch of the river is the Tanis, so that Matthew may have simply replaced an unfamiliar name with one he knew already.[11] This may have been another absent-minded slip of the pen, but we find another hint that he may have had difficulty reading his exemplar. In the Dead Sea, in the list of the five cities of the plain, after the obvious Sodom and Gomorrah and Zoar (*Segor*), the city saved from destruction where Lot took refuge, we have only 'another' and 'another' (*alia, alia*) where there should have been Admah and Zeboyim.[12] Given the general pattern of its shortcomings, this seems more likely to have originated on Matthew's copy than on the map he was copying; it may have been rubbed in places, or Matthew may have had to work against time in a bad light. Certainly we must look to some such explanation for the map's oddest mistake. In the extreme south-east is 'The land of Uz, the country of Job, where were Zeeb, Zebah and Zalmunna'; the names are those of three Midianite rulers, killed on Gideon's conquest of their land,[13] but Matthew has drawn a red rectangle around each, with the letter *c*, turning them into towns.

Even in the entry of real places on the map we may detect further signs of haste. Beside each Matthew has noted in red the sort of place it was, town, castle or village. It seems hardly likely that these notes were taken from wording on the map he was copying; we should suppose rather that his notes correspond to different symbols that were used, like the three symbols used to mark places on the Ashburnham Libri map. The words Matthew used were *ciuitas* for town, *castellum* for castle and *casella* for village, but it was only at Jerusalem, *ciuitas dei*, 'city of God', that he wrote any of these words in full. He seems to have meant *c* to stand always for *ciuitas*, but twice, at Nazareth and Sychar, he put *ciu* instead. *Castellum* appears as *cast'* or *castell'* and once (at Roche Taillée) as *castel*, and *casella* as *cas* or (at two sites south of Jerusalem) as *casel*. That Kerak and Montréal in the south-east are both labelled *c*, presumably for *ciuitas*, must be a further sign of haste – they should surely both have been *castellum*. It is just possible that Matthew's notes of monasteries, mostly *monasterium*, *monest'* or *mon'*, mean that they too were marked by some symbol on his exemplar, but this is less likely; in three places he has *abbacia*, abbey, instead and for two of these, St Symeon and Montferrand, both at the top of the map where he was choosing his colours with care, he used black ink, not the red that he used throughout for *ciuitas*, *castellum*, *casella* and their abbreviations. The original map probably had no special symbol for religious houses.

At least three additions to these notes on places were probably made when Matthew no longer had access to the map he had copied. The neighbouring towns Sebaste and Nablus (entered the wrong way round, as we have seen) had both been marked with a red *c*. Above each of these names Matthew now wrote *ciuitas* in black; beside them he wrote *Idem*, 'The same', with a line drawn to the *c* at Nablus and another drawn to *ciuitas* above Sebaste. This seems to have been a note to himself, a reminder that *ciuitas* is what a red *c* stands for throughout the map.[14] At Tortosa the red *c* for *ciuitas* has been expanded in black to *castrum*, meaning here the same as *castellum*; presumably Matthew learned of the castle there from some other source and the position of Chastel Blanc may well have been altered at the same time. Its name on the coast has been struck out and a note, correctly inland, has been added further

11 Edson, 'Matthew Paris' "other" map', p. 21.

12 Genesis 13.10,11; 14.2,3; 19.22-24.

13 Judges 7.25; 8.21.

14 It is just possible that Matthew's note meant that Nablus and Sebaste were two names for the same place; the so-called Seventh Anonymous account of the Holy Land and the French text 'Les chemins et les pèlerinages de la Terre Sainte' both equate Nablus with Samaria, the alternative name for Sebaste (*Itinera*, ed. De Sandoli, III, p. 82; *Itinéraires*, ed. Michelant and Raynaud, pp. 186, 197). But this was a most unusual mistake, and the lines on the map point clearly to *c* and *ciuitas*, not to the place-names.

south, clearly fitted into space left between the entries of Mount Lebanon and Arqa. Matthew may well have entered correctly the places shown on his exemplar, but they were not necessarily defined or located there with systematic accuracy and precise definition might anyway be difficult – Tortosa, after all, was both town and castle.

To anyone consulting the map, *cas* might as well stand for *castellum* as for *casella*, and *c* might stand for either.[15] These abbreviations show clearly that Matthew did not draw the map for general use; it was for himself, most likely with the intention of making a full copy of his exemplar. Whether he actually did this we cannot tell. Nor can we tell how or where he saw the map he copied, but two possibilities spring to mind. One is that it was at Westminster or some other monastery that Matthew visited. He will have been given the discarded piece of parchment with the 1246 texts already copied on to it, so that he could put them in his chronicle – though, as we have seen, he actually entered there a different version of the letter to the pope. Indeed, Matthew may have specially asked for this sheet, recognising the quality of the pictures on the other side. He then used the blank portion to copy the map before going back to St Albans. In this case the pictures will not have originated at St Albans.

Another possibility is that the map belonged to Henry III, that it was among the furnishings of the itinerant royal household and that Matthew saw and copied it on one of the king's visits to St Albans.[16] It would be fully in keeping with the king's pious aspirations that he should own a detailed representation of the Holy Land and carry it about with him; we know that some fifty years later his son, Edward I, travelled for at least a time with a rolled map of the world, presumably for display at the household's successive stopping-places.[17] These two possibilities are, of course, the merest guesses. But wherever Matthew saw the map that he copied, we may be reasonably sure that it was a sheet map, probably very much larger than his copy. Having to reduce it to fit the space on his parchment would explain some of Matthew's difficulties in its layout.

Matthew had not finished with the piece of parchment when he had completed his map. He used a blank corner (bottom right in relation to the map) to jot down various geographical notes, probably on several occasions. They include an outline itinerary up the Italian peninsula from Sicily to Dalmatia, and some notes broadly relevant to the Holy Land and its neighbouring areas: Tarsus, an archbishopric near Antioch, is where St Paul was born, the balsam vine grows in Idumea, and so on. Reinhold Röhricht included these notes in his typeset reproduction of the map, though they are certainly quite separate from it, and Richard Vaughan, pointing out their value as evidence of Matthew's geographical interests, also provided a transcription.[18] The parchment here is badly rubbed and the notes are not easy to read. It has, however, proved possible to fill in some of the gaps even in Vaughan's transcription and they are reproduced in the appendix to this chapter, p. 73. Even harder to read is what is probably a still later addition to the texts on the parchment, a list of receipts and payments also transcribed here, as far as possible, in the appendix. The list may or may not be in Matthew's hand. And it may or may not have been Matthew who, aware of the quality of the pictures of the Deposition and of the three Marys, placed the bifolium at the start of the Bible he had helped to write, the Bible that is now

15 Thus in the note at Jaffa, 'formerly a town, now a *cas*'' (*quondam c' nunc cas'*), Miller took *cas'* to be *castrum*, and Röhricht (correctly, I think) as some form of *casella* (Miller, *Mappaemundi*, III, p. 154; Röhricht, 'Karten und Pläne VI', p. 179). Either word might be seen as fitting the facts.

16 Vaughan, *Matthew Paris*, pp. 12-13, lists the nine occasions when the king visited St Albans between 1220 and 1259.

17 P.D.A. Harvey, 'Maps of the world in the medieval English royal Wardrobe', *Foundations of medieval scholarship: records edited in honour of David Crook*, Borthwick Texts and Studies 36 (York, 2008), p. 52.

18 Vaughan, *Matthew Paris*, p. 240.

Corpus Christi College MS 2. It was presumably at this point that the parchment was cropped to fit the volume. Perhaps a couple of millimetres were taken from along the top, and probably about a centimetre from each side, with the loss, in the Mediterranean, of probably 'MEDITERRANEUM' and certainly parts of 'MARE' and Tripoli's peninsula and, on the other side, of parts of the geographical and financial notes.

The first printed reference to the map is in the catalogue of the college's manuscripts that was published in 1852: 'A geographical picture covering Antioch, Syria and Palestine, with explanations'.[19] It was probably this entry that drew it to the attention of Konrad Miller and Reinhold Röhricht, who both, independently, published accounts of it in 1895. Röhricht, in his series of articles on medieval maps of Palestine, gave a two-page description and a drawn reproduction of the whole map, with typeset wording, and wrote that the author 'was certainly an Italian'.[20] Miller, who included regional maps in his six-volume work on medieval world maps, transcribed the map's wording, identifying locations and sources, but only from the left-hand page of the bifolium; probably he had been sent a photograph only of this page and assumed it was the complete map. In an impressively perceptive note, however, he pointed out that the hand of the map bears a 'startling similarity' to that of Matthew Paris's maps of Britain and that the map has points of contact with Matthew's Acre map of Palestine; he saw the map as work either of Matthew himself or of one of his pupils.[21] Photographs of the left-hand page, much reduced, were published in 1936 by Prince Youssouf Kamal in his vast collection of early maps of Africa and Egypt and in 1958 by Richard Vaughan in his book on Matthew Paris.[22] Vaughan accepted that the whole map was in Matthew's hand – in an earlier article he had suggested that parts were not – while doubting whether it was his own original work; in describing it he underlined its great interest.[23] Thus fully established as drawn by Matthew, the map – or its left-hand page – has been several times reproduced in colour and has been briefly discussed by Evelyn Edson in 1994 and by the present author in 2001.[24]

The map's framework, including probably most towns and other major features, was taken from a map of the world.[25] The second Tournai map of Palestine likewise took from a world map a completely straight coastline from St Symeon to Cairo. The peninsula at Tyre is a fact of geography known to medieval mapmakers and the peninsula at Tripoli may be an echo of the peninsula at Sidon on the Ebstorf world map and the Ashburnham Libri map. But we see the map's debt to contemporary world maps more clearly if we look at some of its details. The sea of Galilee is drawn on the map with the line of the river Jordan passing straight through it, as though the lake's shores had been added as an afterthought. Given the constraints of the page, and the difficulties Matthew had got into when he copied this part of the map, we might reasonably suppose that this was exactly what happened. However, we see just the same feature on two of the three versions of Matthew's Acre map of Palestine, both carefully drawn as fair copies, though here it is the Dead Sea that the river passes through; the sea of Galilee is not shown. A river that enters and leaves a lake would no more be shown like this on a medieval map than on a modern one – yet we find the same feature in the sea of Galilee on the Sawley world map of about 1200 **[38]**.

19 Coxe, *Catalogus codicum*, II, Corpus Christi College, p. 1 ('Tabula geographica Antiocham, Syriam et Palaestinam complectens, cum explanationibus').

20 Röhricht, 'Karten und Pläne VI', pp. 178-82, plate VI; 'der sicher ein Italiener war' (p. 180).

21 Miller, *Mappaemundi*, III, pp. 152-6 (the photograph promised on p. 152 was not in fact published); 'auffallende Aehnlichkeit' (p. 156).

22 Kamal, *Monumenta cartographica*, IV, fasc. 1, f. 112; Vaughan, *Matthew Paris*, plate XVII.

23 Vaughan, 'Handwriting of Matthew Paris', p. 391; Vaughan, *Matthew Paris*, pp. 245-7, where he mistakenly sees the map as a new discovery.

24 Nebenzahl, *Maps of the Holy Land*, p. 39; Edson, 'Matthew Paris' "other" map'; Harvey, 'Paris's maps of Palestine' (the present chapter is an expanded version of its account of the Oxford map); *Map book*, p. 55.

25 Alberto Capacci, in *Cristoforo Colombo*, ed. Cavallo, p. 466, suggests that the places shown on the Ashburnham Libri map with a gold symbol correspond to those marked *ciuitas* on the Matthew Paris Oxford map, the other places to those marked *casale* (i.e., more likely, *casella*). Certainly none of the places marked with a gold symbol is defined as anything other than *ciuitas* on the later map (except the special case of Jaffa), but this is only fourteen out of thirty-two gold symbols, and all are anyway likely to be called towns: Acre, Beirut, Nazareth, etc. Of the thirty other settlements named on the Ashburnham Libri map only one is called *casella* (*Betnoble*, i.e. Nob) on the Matthew Paris Oxford map and two are called *ciuitas* (Kerak and Montréal, whether or not correctly). I cannot see any basis for the relationship Capacci suggests.

38. Lines of the river Jordan on the sea of Galilee and the Dead Sea. Unusually, the river's lines continue uninterrupted across the sea of Galilee on the Sawley world map (top left) and on Matthew Paris's Oxford map (below) and across the Dead Sea on his Acre map (top right). This was probably a detail from the general pool of precedents drawn on for world and regional maps. Cambridge, Corpus Christi College, MS 66, p. 2; Oxford, Corpus Christi College, MS 2*, f. 1v; BL, Royal MS 14 C.vii, f. 5r.

It may have originated simply by accident, but may perhaps stem from Adomnan's account of the way the stream of the Jordan could be distinguished by its milky colour 'for a considerable stretch' after it had entered the Dead Sea.[26] This is not to suggest that the world map that contributed to the map Matthew copied was itself directly related to the Sawley map; it is rather that this detail came from a general pool of detailed representations on world maps that a mapmaker who was composing a new one would draw on eclectically, copying this feature from one map, that from another.

We see the same process at work in another detail: the bird that is perched on Mount Gilboa. A note tells us that 'Here Saul fell' (*Hic cecidit Saul*) and this was indeed where the Philistines defeated the Israelites and killed Saul.[27] There is nothing in the Bible to show why a bird should have been drawn there, yet we again see a bird on Mount Gilboa on the Vercelli world map. Moreover, on the Hereford world map we see a very similar bird perched on a hill. The hill is not Mount Gilboa, but is not far away, beyond the south end of the Dead Sea, and the bird is identified as the phoenix **[39]**. Again we should assume no direct connection between the map that contributed to Matthew's exemplar and the Vercelli or Hereford map, but it is likely that it is the same bird, copied again from one world map to another, and capable of playing a different role if called upon to do so.

There are other details that probably reached the map from a world map drawn on by Matthew's exemplar. The figure of Lot's wife turned to salt can be seen also on the Hereford map, while the star at Bethlehem and the jars of water turned to wine at Cana appear on the Ebstorf map. In these cases we might well suppose that Matthew,

26 *Adamnan*, ed. Meehan, pp. 88–9.
27 I Samuel 31.8.

39. Bird on Mount Gilboa. For no obvious reason a bird is perched on Mount Gilboa on Matthew Paris's Oxford map (left) and on the Vercelli world map (centre), and both resemble the Hereford world map's not-far-distant phoenix (right); probably all three stem ultimately from the same model. Oxford, Corpus Christi College, MS 2*, f. 1v; Vercelli, Biblioteca capitolare; Hereford, Mappa Mundi Trust.

well used to bringing graphics into his chronicles, had added these pictorial details on his own initiative, and indeed if what we had was a fair copy of the map we might hardly have doubted that this was how they got there. But there is no other sign on the map that Matthew introduced flourishes of his own, whether in his hurried copying or afterwards. Lot's wife, the star and the jars of wine most likely figured on Matthew's exemplar, having originated in a world map.

To the framework taken from a map of the world, Matthew's exemplar added further information from a description of the Holy Land. It can be identified, as Miller pointed out in 1895, as one written in French that circulated in varying versions in the early and mid-thirteenth century.[28] In 1882 Henri Michelant and Gaston Raynaud printed versions from four manuscripts, with variants from a fifth; they presented them as two different texts, one *Les pelerinaiges por aler en Iherusalem* of about 1231, the other, *Les chemins e les pelerinages de la terre sainte*, dating from before 1265. However, Comte Riant, in his introduction to their edition, makes it plain that this is an arbitrary distinction and that the dating is problematic: what we have is a single text that survives in rather varied versions.[29] Clearly it was a much copied work that probably survives in many more manuscripts than were known to Michelant and Raynaud.

The map underlines this. Most of the features in its central region, broadly the area bounded by Acre, Ascalon, Hebron, Jericho and Damascus, appear in this text, and there can be no doubt that the two are closely linked. Indeed, we can see the map as a graphic representation of the text, but not precisely of any one of the published versions. Thus it includes Dok and Recordane, and Abraham's garden near Jericho, which all appear in only one of the five manuscripts,[30] the site of the angels' appearance to the shepherds, *Gloria in excelsis*, which appears only in another,[31] and the church of Samuel, mentioned only in the three remaining manuscripts.[32] Presumably the map used a version of the text that included all these, while omitting other sites that appear in the published versions: by no means all the places they mention can be found on the map. The text gives many indications of direction and distance and there seems no reason why its information could not have been entered directly on Matthew's exemplar, within the framework provided by the world map.

Between them these two sources – the world map and the written description – account for some two-thirds of the features and notes on the map. But the mapmaker must have drawn on other sources too. The world map will have shown what places there were, but not whether they were towns, castles or villages. Much of the

28 Miller, *Mappaemundi*, III, pp. 155-6.

29 *Itinéraires*, ed. Michelant and Raynaud, pp. xix-xxi, xxvii-xxviii, 87-104^{7}, 177-99.

30 Ibid., pp. 196, 198.

31 Ibid., p. 104^{6}.

32 Ibid., pp. 93, 181.

information in the north-west will have come from some other source. Few of the contemporary accounts of the Holy Land that have been published give distances between places in day's journeys, and none are the same as those on the map. Thus Othmar has one day from Tripoli to Byblos (the map has two days), then one day to Beirut (also one day), one day to Sidon (two days), two days to Tyre (one day) and two days to Acre (one day); Wilbrand of Oldenburg has one day from Tortosa to Tripoli (the map five days).[33] The map's 6 leagues from Acre to Haifa are found also in a French version of the work of Antoninus of Piacenza,[34] but other descriptions have a shorter distance: 4 or 3 leagues, or 10, 8 or 4 miles.[35] Beside the pilgrim route from Jaffa to Jerusalem is the note 'They go as pilgrims, they come back as palmers' (*Eunt peregrini, redeunt palmarii*), that is, bearing palms as a symbol of their pilgrimage; this may be ironic, not simply factual, for palmers were noted for soliciting alms.[36]

The map may well include information obtained at first hand from a crusader or other traveller, but it was probably created in western Europe: the world map that was its starting point would most likely be seen in northern France or England. The date of its creation is difficult to establish; the map that Matthew copied may itself have been a copy, even at more than one remove from the original, and details may have been added or left out each time it was copied. Miller supposed it must date from after 1250, but apparently only from an impressionistic view of its spellings of place-names.[37] Evelyn Edson suggests a date after 1235, when a dispute arose over water rights at Dok and Recordane; these names probably reached the map from the French text, but this refers to the mills at Dok and the dispute could well be why they are mentioned there.[38] She also points to the map's omission of Beaufort and Safed, castles regained by the Templars in 1241, and Röhricht points to its omission of the castle at Montfort, under construction in 1230.[39] Röhricht also points out that the map's pilgrim route from Jaffa to Jerusalem suggests a date between 1229 and 1239, when Jerusalem was in Christian hands; the route was placed under Christian control by the agreement of 1229.[40] Some of these indications are contradictory, none is hard evidence and the map's omissions are anyway a most unreliable guide, but taken together they seem to point to a date around 1230.

The most remarkable feature of the map is that it has north at the top. This can be paralleled by the ninth-century map of the Land of Promise and by some of Rashi's maps of Canaan and their derivatives,[41] but the world map with which the mapmaker began his work will surely have been oriented to the east. He may have changed its orientation to achieve upright format in a sheet map intended for display. Transposing the map in this way will have involved much careful work and so too, if the interpretation offered here is correct, will the conversion of the probably undifferentiated places shown on the world map into the appropriate pictures or symbols for towns, castles and villages. We start to see the regional map's creator as imaginative and ingenious, and his map a worthy subject for Matthew to copy. We may regret that it has not survived; and, if Matthew produced a fair copy from his draft, that this has not survived either. But we must be grateful to Matthew for giving us this glimpse of what must have been an extraordinarily interesting and handsome map.

33 *Itinera*, ed. De Sandoli, II, p. 302; III, p. 210. Byblos is *Gibel* in Othmar's text, *Iubeleth* on Matthew's map.

34 *Itinera hierosolymitana et descriptiones terrae sanctae bellis sacris anteriora et latina lingua exarata*, ed. by Titus Tobler and Augustus Molinier, Publications de la Société de l'Orient latin, série géographique 1,2 (2 vols, Geneva, 1879), I, p. 383.

35 *Itinera*, ed. De Sandoli, I, p. 318; III, pp. 232, 304, 450, 468; IV, pp. 58, 202, 246, 264.

36 Unsurprisingly, this note defeated both Röhricht, 'Karten und Pläne VI', plate VI, and Miller, *Mappaemundi*, III, p. 154; the *p* of *palmarii* is all but invisible as it is written on the brown frame of the picture showing through from the other side of the leaf.

37 Miller, *Mappaemundi*, III, p. 156.

38 Edson, 'Matthew Paris' "other" map', p. 21; *Itinéraires*, ed. Michelant and Raynaud, p. 198.

39 Edson, 'Matthew Paris' "other" map', p. 21; Röhricht, 'Karten und Pläne VI', p. 179.

40 Röhricht, 'Karten und Pläne VI', p. 179; above, p. 4.

41 Above, pp. 19-22.

APPENDIX

Expansions of abbreviations are in italics; the number of stops in square brackets corresponds to the likely number of illegible letters. In the accounting notes li', s', d', ob' (pounds, shillings, pence, halfpenny) have not been expanded.

1 GEOGRAPHICAL NOTES

¶ [. . .]sana p*ro*pinq*u*ior *est* terre s*an*cte q*uam* Bru*n*dusiu*m* q*uia* Nauigantib*us* a Massilia i*n* terra*m* s*an*cta*m* *est* messina media g[. . .]a ¶ Marsilia *est* co*nter*mina hyspanye

¶ Silicia*m*

¶ apulia*m*

¶ Calabria*m*

¶ [. . . . .]

¶ T*er*ra*m* imp[eratoris] q*ue* durat vsq*ue* ad alpes

¶ Ca*m*pania*m*

¶ Roma*m* in ca*m*pania

¶ Valle*m* spolet[i]

¶ Valle*m* anconie

¶ Venicia*m*

¶ Dalmacia*m*

¶ i*n* arabia *est* ydumea ubi crescit vinea balsami

¶ [in] alia fine est Sabea p*at*ria [r]egi*n*[e et] media co*nter*mine su*n*t indie

[g]alathia i*d est* t*ur*chia

pa*m*firia
armenia } Id*em*

[R]ex arrag*one* adept*us est* s*uper* sarracenos
i*n* hisp*ania* xxx dietas

[C]atalonia *est* p*at*ria co*nter*mi*na* p*ro*ui*n*cie

[V]all*is* Moriant ¶ Sabaudia

Tarsus *est* archiep*iscopatus* prope antiochia*m* ubi nat*us est* S*anctu*s paul*us et est* i*n* cilic*i*a armenie minoris

i*n* p*ar*te boriali *est* Ruscia *et* [.]amania *et* Blakania *et* s*uperius* u*ersus* anthiok' *est* yconiu*m*

corsa *est* i*n*sula p*ro*pe Ianua*m* ubi optimi s*un*t ancipitres

[. . . . .] i*n*sula u*bi* exulau*it* [. . .] *ue*l [. . . . .] Ioue*m* seu [. . . .] prior

2 ACCOUNTING NOTES

Sum*m*a Den[ariorum]

De Ric' [. . . . .]

[. . . . .]end' suu*m* xiij d'

P*ro* sale [. . . . .] d'

I*n* [. . . . .] x d'

I*n* expe*nsis* ad lond' vj d'

P*ro* [. . . . .] d'

P*ro* [. . . . . o]b'

It*em* p*ro* bob*us* ad [. . . .] x d'

Debent x li' [. . . . .] xiii d'

I*n* expensis x[ij] d'

W[. . . . .]

8 THE ACRE MAP OF MATTHEW PARIS
mid-13th century[1]

B[2] London, British Library, Royal MS 14 C.vii, ff. 4v-5r **[40]**. Parchment, 34.5 × approximately 49 cm. Dark brown, blue, green and red ink, with light brown, green, grey and red paint.

C Cambridge, Corpus Christi College, MS 26, ff. iii v-iv r **[41]**. Parchment, 36 × 54 cm. Dark brown, blue and red ink, with blue, light brown, green and grey paint.

D Cambridge, Corpus Christi College, MS 16, ff. iii v-iv r (until 2003, ff. ii v, v r) **[42]**. Parchment, 23 × 49 cm (lower part of each page lacking). Dark brown, blue and red ink, with blue, light brown, green and grey paint.

MATTHEW PARIS'S PRINCIPAL HISTORICAL WORK was the *Chronica majora* (*Greater Chronicles*), 'the most comprehensive history yet written in England'.[3] It is a history of the world from the Creation onwards and, while it concentrates increasingly on English affairs as it approaches the mid-thirteenth century, it still describes many events elsewhere in Europe. The detailed, careful analysis of the manuscript text by Richard Vaughan has unravelled the complexities of its composition and has shown that all the later parts of the work are in Matthew's own hand. By the time he died in 1259 the text had spread into three volumes. Two are now in the Parker Library at Corpus Christi College, Cambridge. The first, MS 26, contains the history down to 1188, continued in the second, MS 16, to 1253. The third volume, now Royal MS 14 C.vii in the British Library, first contained only Matthew's *Historia anglorum* (*History of the English*), an abbreviated version of the *Chronica majora* covering English affairs from 1066 to 1253, also written in his own hand; however, he supplemented this with a continuation of the *Chronica majora* from 1254 to the time of his death, an event recorded in the manuscript by a modestly anonymous fellow-monk, with a picture of Matthew on his death-bed.[4]

Each of these three volumes contains one of Matthew's three maps of Palestine that are dominated by the plan of Acre; the three are not identical, but are different versions of the same map. It should be said at once that while the chronicle naturally refers to some places shown on the map – Jerusalem, Acre, Cairo and others – there is no detailed correspondence; the map was not drawn to show the chronicle's readers where the places it mentions lay. In each case the map was included in what was probably a single quire of illustrative matter placed at the front of the volume; most of the material is in Matthew's hand and it was presumably he who prefaced his work in this way. In MS 26 two further illustrative leaves at the end of the text were probably originally at the front and some leaves have been lost, others cut down, in the first quire of MS 16. However, these preliminary pages were never uniform in all

1 I am most grateful to Professor Denys Pringle for his knowledgeable advice on many points that arose in this chapter, which is an expanded and corrected version of the sections dealing with the Acre map in Harvey, 'Paris's maps of Palestine'.

2 I follow here the letters used by both *Itinéraires*, ed. Michelant and Raynaud, p. 124, and Miller, *Mappaemundi*, III, pp. 90-1; their A is the early-modern copy of B in BL Lansdowne MS 253, ff. 194v, 197r (below, pp. 91-2). I do so with hesitation, as they differ, confusingly, from the letters used by Vaughan, *Matthew Paris*, p. 241. In Harvey, 'Paris's maps of Palestine', to avoid confusion, I refer to the first-, second- and third-volume maps (i.e. the volumes of the *Chronica majora*) but this is too clumsy to be used for the more extended discussion here.

3 Gransden, *Historical writing*, p. 359.

4 Vaughan, *Matthew Paris*, pp. 35-77, 110-58; Gransden, *Historical writing*, pp. 356-74.

three volumes. They include some of Matthew's best work as an artist: a picture of himself kneeling before the Virgin and Child in MS 14 C.vii and heads of the Virgin and Child, of Christ living and of the crucified Christ in MS 26. There are Easter tables and a calendar in both MS 14 C.vii and MS 26. In MS 16 there is a long description with a picture, drawn from life, of the elephant that King Louis IX of France gave to King Henry III and in MS 14 C.vii there are portraits of the kings of England from William I to Henry III, each holding a miniature church. There are three other maps: a world map in MS 26 and maps of Britain in MS 14 C.vii and MS 16.[5]

In all three volumes there is a diagram of winds, but otherwise the only prefatory items that appear in all three volumes are the map of Palestine and, immediately preceding it, an illustrated itinerary from London to south Italy. It seems that in each case this itinerary occupied five pages; the map then extended across the back of the fifth page and the facing page, and in MS 14 C.vii and MS 16 this was a bifolium in the centre of the quire. Besides being physically linked in this way, the itinerary and map have been seen as a single work, the pilgrim route from London to Jerusalem, and it is likely that this is how Matthew himself envisaged them. It is interesting that in MS 26 parts of the Palestine map on the verso of f. iii were demonstrably drawn before parts of the itinerary's last page on the recto.[6] There is certainly a fourth copy of the itinerary, in Matthew's hand, without the Palestine map, but it is simpler in form, on only two pages, a single opening in another St Albans manuscript, and we may see it as an early draft or an abridgement.[7] At the very least the itinerary provides the context for the map of Palestine, and Katharine Breen argues that the map of Britain that follows the map in MS 14 C.vii should be seen as part of the sequence, the pilgrim's return home;[8] there is no knowing whether or not this was in Matthew's mind, but in MS 16, though not in MS 26, a map of Britain again follows the map of Palestine.

The itinerary is a quite extraordinary work, arguably the most interesting of all Matthew's cartographic productions: a strip-map with pictures representing the towns *en route*, mostly one day's journey apart. Rivers and their crossings are shown, and for parts of the distance alternative routes are offered. While we are used to this form of road map – though mostly without the pictures – there is no known contemporary or earlier medieval parallel that shows a route in this way and it is not surprising that it has attracted scholars' particular attention. In 1999 Daniel K. Connolly pointed out that in concept it is akin to the Peutinger map, the late-Roman route-map that survives in a twelfth-century copy.[9] Since then, Patrick Gautier Dalché has shown that another copy was at Padua in the fifteenth century and Richard Talbert has raised the possibility that there may also have been a copy in thirteenth-century England, providing the idea, if not precisely the model, for the Gough map of Britain with its network of roads.[10] In this case the same copy of the Roman route-map may equally have given Matthew the idea of giving graphic form to a written itinerary – but this is advancing far into speculation, and we really need look no further for Matthew's inspiration than his own inventive and imaginative mind. We may even see Matthew's graphic itinerary developing in the way the four successive versions of his map of Britain show the route from Dover to north-east

5 Warner and Gilson, *British Museum: catalogue of Royal manuscripts*, II, p. 135; James, *Manuscripts in Corpus Christi College, Cambridge*, I, pp. 51, 53-5.

6 While the work was in progress a triangular piece of parchment, some 6 by 11 centimetres, was torn and cut from the bottom outer corner of f. iii and replaced by a patch glued on (f. ii suffered similar damage and repair). This part of the verso had already been drawn, for the green of the patch is a different shade from the surround; on the recto, however, the vertical bar, Arezzo and the man with a mule were drawn after the patch had been added **[41, 43]**. The vertical bar's failure to join up must result from differing expansion or contraction of the parchment in page and patch.

7 The itinerary is described, illustrated and discussed by Vaughan, *Matthew Paris*, pp. 247-50, plates XII, XIII, XV; Lewis, *Art of Matthew Paris*, pp. 323-47; Danielle Lecoq, 'Mathieu Paris: de l'itinéraire à la représentation du monde', in *L'oeil du cartographe*, ed. by C. Bosquet-Bressolier (Paris, 1995), pp. 20-7; Connolly, 'Imagined pilgrimage'; Connolly, *Maps of Matthew Paris*; Sansone, *Mattheus Paris*, which includes full colour plates of the versions of the itinerary. The draft or abridged version of the itinerary is BL Cotton MS Nero D.i, ff. 183v-184r; it is reproduced in Connolly, *Maps of Matthew Paris*, fig. 10.

8 Katharine Breen, 'Returning home from Jerusalem: Matthew Paris's first map of Britain in its manuscript context', *Representations*, 89 (2005), pp. 83-7.

9 Above, pp. 17-19.

10 Connolly, 'Imagined pilgrimage', p. 619, n.17; Patrick Gautier Dalché, 'Du nouveau sur la transmission et la découverte de la Tabula Peutingeriana: la "Cosmographia vetustissima" de Pellegrino Priscini (†1518)', *Geographia Antiqua*, 13 (2004), pp. 71-84, plates I-IV; Talbert, *Rome's world*, pp. 170-2. Sansone, *Mattheus Paris*, p. 94, also postulates a link between Matthew's itinerary and the Peutinger map.

DAMAS
Saphat
la tur de pisanz
LA VILE DE ACRE
le chastel le rei de acre
la porte pardeus
la maisun del hospital seit Johā
le tem ple

40. The Acre map of Matthew Paris, map B, mid-13th century; east at the top. The most carefully executed of our three copies of the map, and very likely the first to be drawn. Its inscriptions are shorter than in the other copies and a higher proportion are in Latin, not French. BL, Royal MS 14 C.vii, ff. 4v-5r.

England, the backbone of each map. On what is probably the oldest of the four a line is drawn between the name of each town, with a pictorial symbol only at London; on what is probably the latest, and certainly the most finished, version each town has a symbol beside its name, though the lines joining them have gone.[11]

11 *Four maps of Great Britain*, ed. Gilson; Harvey, 'Paris's maps of Britain', pp. 114–18.

It is the last page of the itinerary, where it reaches Rome and south Italy, that most

41. The Acre map of Matthew Paris, map C, mid-13th century; east at the top. This has more text than our other two copies of the map, and may well have been the last to be drawn. On the left-hand page the map continues on a parchment extension sewn on to the leaf. Cambridge, Corpus Christi College, MS 26, ff. iii v-iv r.

closely concerns the map of Palestine drawn (its left-hand side) on the back of the leaf. In MS 16 this is the only page of the itinerary that survives, and this only in part: the bottom part of the whole bifolium has been cut away so that we are left with only the top two-thirds of the end of the itinerary, of the map of Palestine and of the map of Britain that is on the fourth page. In the other two manuscripts the itinerary and map

are complete. However, even here, in its last page the itinerary is 'very muddled and confusing'[12] – and it is not even consistent between the three manuscripts. We need not follow the intricacies of its course, but two of its features affect the map of Palestine on the back of the page.

One is the way it shows Sicily. When the route had reached the end of the toe of Italy Matthew had reached the top of the page. Resourcefully, he added there a small piece of parchment, more or less triangular, attached by a parchment strip threaded through slits in the page (MS 14 C.vii) or sewn with thread (MS 26).[13] On this he drew a map of Sicily with its mountains and towns. But this small piece of parchment could be folded downwards as a flap, and on the back Matthew wrote a geographical note about Sicily with a little picture of Etna. When the flap is folded out it appears

12 Lewis, *Art of Matthew Paris*, p. 340.

13 It is now glued and sewn with modern thread but presumably through original sewing holes.

above the map of Palestine, its note on Sicily upside down. This extra piece of parchment for Sicily is missing from MS 16, but we can see the slits near the top of the page for the parchment strip that once held it in place. Rome, unlike Sicily, appears in the itinerary, but in MS 14 C.vii a similar but rectangular piece of parchment is attached to the right-hand side – the outside – of the page, with a stylised plan of Rome, a battlemented wall around it and the river Tiber flowing through the middle. On the back – again it can be turned in as a flap – is a short historical note on the city. Like the attachment for Sicily it is missing from MS 16, but two short notes on the page refer to it and there is indeed a series of sewing holes and a fragment of thread on the outer edge of the page. On MS 26, however, the parchment extension for Rome takes an altogether more ambitious form. A piece of parchment

42. The Acre map of Matthew Paris, map D, mid-13th century; east at the top. The lower half of each leaf has been cut away. On what remains, there is more text than on map B, less than on map C. Cambridge, Corpus Christi College, MS 16, ff. iii v-iv r.

10 centimetres wide has been sewn on to the whole height of the page, and on it is the plan of Rome with a long historical note **[43]**. It was not, however, designed to be turned over as a flap, and the other side simply extends the map of Palestine for an extra 10 centimetres, giving added space for its many inscriptions.

The extension on the left side of map C, the map in MS 26, is only one of many differences between the three versions of the map of Palestine. But they are alike in their basic structure – though we must remember that we have only the top two-thirds of map D in MS 16. East is at the top and the river Jordan, with the Dead Sea and its sources the Jor and the Dan (but not the sea of Galilee), appears at the very top of the page on the right, Damascus at the very top of the page on the left. To the south the map extends to the river Nile, with Alexandria and Cairo. To the north are the Caspian Mountains, Antioch and, on the coast, Tyre. Along the coast to the south is the plan of Acre, hugely out of proportion, occupying a right-angled turn in the coastline. Further south the coast continues with a series of towns from Haifa to Damietta and the Nile, while on the sea are ships, some with sails and some crowded with pilgrims – or crusaders, for those in one boat in MS 26 are armed with assorted weapons. Jerusalem and a few other inland towns and hills are shown, but most of the spare space on the map is taken up with blocks of writing, some of them single sentences, others long paragraphs.

All this is common to all three versions of the map – always with the proviso that part of D is lost – and so too are most features of the plan of Acre. This plan must have been Matthew's starting-point for the whole map, and while we rightly see it as a map of Palestine, it is not far removed from some twelfth-century plans of Jerusalem that set the city in context by showing other places outside and around it, as far distant as the Jordan and the Mediterranean.[14]

All three maps show Acre as a walled city. On B the walls are battlemented and surround the whole inner area, on what we can see on D they are marked only by a single line, and on C they are again battlemented but appear only on the landward side – there is no wall between the city and the shore. On the north side of the city is a further enclosed quadrant, the suburb of Montmusard that was developed in the twelfth century, shown with a wall on its landward side but not along the shore.[15] On B, just possibly significantly, battlements have been added to it as an afterthought, after the adjacent red inscription had been entered and partly in the same red ink; on D, as on the main city wall, there are no battlements, while on C the wall was drawn fully battlemented from the start. East of Montmusard a circular enclosure, identified as the cemetery of St Nicholas, balloons out from the city wall; its pictures of tombs are not further named on B, but on D they are called *charner* and on C *sepulchres*.

On the eastern wall all these maps show the Accursed Tower (*la tur maudite*), a gate leading to the mill at Dok and, at Montmusard, St Nicholas's gate which on C gives directly on to the cemetery. On B two further gates, unnamed, are drawn at the north-west corners of the city wall and of Montmusard. Within the walls no streets are shown, only a selection of major buildings; there is a small picture of each and nearly all are named. The three in the surviving part of D are the same as on the other two maps, but comparison of those on B and C is harder than one might suppose. On each, the centre fold of the bifolium is a little way in from the south shore of Acre.

14 E.g. BL Harley MS 658, f. 39v (above, pp. 23-4).

15 D. Jacoby, 'Montmusard, suburb of crusader Acre: the first stage of its development', *Outremer: studies in the history of the crusading kingdom of Jerusalem presented to Joshua Prawer*, ed. by B.Z. Kedar et al. (Jerusalem, 1982), pp. 205-17; Pringle, *Churches*, IV, pp. 152-3

On B the two leaves of the bifolium have been badly damaged on the gutter and parchment has been lost – probably between 10 and 15 millimetres – along the whole height of the fold and one, even two, buildings may be lost without trace.[16]

It is not impossible that exactly the same buildings were meant to appear on both B and C – indeed, on all three maps – though the tower of the Pisans appears on B but not on C, the house of the patriarch on C but not on B. The house of St Thomas of Canterbury is named only on C, but on B there is an unnamed building in the same position. On both maps a name is duplicated, given to two separate buildings: the hospital of the Germans on B (it is named once on C) and the tower of the Genoese on C (it is named once on B). The other buildings on both maps are the castle of the king of Acre, the house of the Hospitallers, the house of the Knights of St Lazarus, the house of the Templars and the house of the constable. In the south-west corner on B, close to the area lost, is *la chemin sus mer*, the terminal point of the sea-route to the Holy Land, and on C *la chaene*, the name of the court dealing with maritime affairs.[17]

We have no other plan of Acre from the thirteenth century, but from the first half of the fourteenth we have two, in the same group of maps as the grid maps abridged from the large Burchard map of Palestine (chapter 10, pp. 109-10, 113-14). One is found in six manuscripts of Marino Sanudo's *Liber secretorum fidelium crucis* and in a separate volume of the maps from this work, books that are probably all to be dated between 1320 and 1331, as well as in a fifteenth-century copy; it probably originated in the workshop of Pietro Vesconte, at Venice, where Sanudo's work was copied. The other, wholly independent of the plan in Sanudo's work, is in three manuscripts of the *Chronologia magna* by Paolino Veneto, bishop of Pozzuoli, one copied soon after 1323, another in 1329 or a little later and the third between 1334 and 1339; the three are not identical. Both are more detailed than Matthew's plan of Acre; each names some fifty local features, with an intricate pattern of streets.[18] Neither is related to Matthew's plan – there is no common source – and Matthew probably compiled his plan himself. It is not so complicated that it could not have been drawn from the description – and with the help – of someone who knew Acre well. This might well have been the master of the foundation of St Thomas of Canterbury at Acre, who visited St Albans in 1257, but Matthew will certainly have met others too who could have supplied the information.[19]

We cannot rely on Matthew's plan as evidence of Acre's topography; it is significantly less informative than the fourteenth-century plans. One feature that Matthew misplaced is the tower of the Pisans, shown in the north of the city whereas it actually lay in the south.[20] But Matthew's first priority will have been to place features in the right part of the map, in this case within the walls of Acre; he will have been less concerned with their relative positions there and will have placed the tower of the Pisans simply in a convenient blank space. In one of his maps of Britain, entering a list of towns along the south coast, he ran out of space and intercalated the remaining names between those already there; putting them in the right part of the map was what really mattered, and getting them in the right order was less important.[21] In fact, Matthew's plan of Acre was a considerable achievement, imaginative and innovative. We have no comparable plan of any town drawn in England in the thirteenth century or earlier.

16 To re-create the bifolium, a parchment strip was at some point attached to the inside edges of the two leaves and on both sides of f. 5 this strip was painted green, slightly disguising the edge of the original parchment. Other leaves of the quire were similarly treated. In 1952, ff. 2-5 (i.e. the itinerary and the maps of Palestine and Britain) were detached from the volume to be preserved separately (note by A.J. Collins, then Keeper of Manuscripts, at the front of the volume). Until MS 26 was rebound in 2003 its tight binding meant that a similar strip in the middle of C was also lost to sight in the gutter; now, however, it is fully visible.

17 Jonathan Riley-Smith, *The feudal nobility and the kingdom of Jerusalem* (London, 1973), pp. 91-2.

18 Röhricht, 'Marino Sanudo', pp. 124-6; Degenhart and Schmitt, 'Sanudo und Veneto', pp. 22-5, 78-80, 105, 120-1; Jacoby, 'Crusader Acre'. Besides those entered by Degenhart and Schmitt in the table on p. 105 there is a plan of Acre in Paris, Bibliothèque nationale de France, ms latin 4939, f. 113v. Matthew's and the 14th-century maps are all reproduced, and their wording transcribed, by B. Dichter, *The maps of Acre: an historical cartography* (Acre, 1973), pp. 9-30, while the details of Acre's topography, with close reference to the maps, are fully discussed by Pringle, *Churches*, pp. 3-35 and index references under 'Matthew Paris', 'Paolino Veneto' and 'Vesconte'.

19 *Matthaei Parisiensis chronica majora*, ed. Luard, v, pp. 630-1.

20 Jacoby, 'Crusader Acre', pp. 6, 11, 25-6.

21 Harvey, 'Paris's maps of Britain', p. 117.

43. Itinerary from London to Apulia by Matthew Paris, mid-13th century. The last section, with an account of Rome on the parchment extension sewn to the right of the page and a map and note on Sicily on the triangular addition at the top. The north half of map C of Palestine is on the other side of the leaf and the triangular patch at the foot of the page is visible here and in **[41]**. Cambridge, Corpus Christi College, MS 26, f. iii r.

It is likely, though, that Matthew had not only seen one of the many copies of a highly stylised plan of Jerusalem that circulated in the twelfth and thirteenth centuries, but actually had it by him while he drew the successive versions of his map.[22] On all three Jerusalem appears as a battlemented square, far smaller than Acre, and within it, in the same positions as on the plan, are pictured and named the Temple of the Lord, the Temple of Solomon[23] and the Holy Sepulchre.[24] Outside the wall the valley of Jehoshaphat is named as the burial place of the Virgin Mary. On C there are no other local names, but B and D both show the mount of Olives, B shows the pool of Siloam and D names the tower of David, the gate on to the road to Jaffa and the Golden Gate (*porte sorres*).[25] Some of these names appear on world maps, but we have no contemporary world map that shows the holy sites within the walls. Matthew's source was almost certainly a plan of Jerusalem.

It may seem surprising, then, that Matthew drew Jerusalem as a square, not as the circle that appears on most of the contemporary plans – and on the Hereford world map. On the Ebstorf world map we see a square Jerusalem, but this is the explicitly square new Jerusalem of the book of Revelation, for its twelve gates are clearly shown.[26] It is just possible that Matthew had this symbolism in mind,[27] but it seems unlikely, for the buildings, the gates, all the topographical detail in and around the city, relate firmly to Jerusalem on earth. So too do other notes within the walls on B and C: 'The city the most exalted of all cities' on B and a more extended version of the same on C. On D is a note more ambiguous in its connotations, 'The vision of peace' (*visio pacis*) beside a blazing sun or star; but this is simply an interpretation of the name 'Jerusalem' that might apply to either city, earthly or heavenly. Contemporary Jerusalem was in fact roughly quadrilateral and one twelfth-century plan shows it thus with a street pattern and other features clearly drawn from life. Besides this, one fourteenth-century copy of the stylised plan that Matthew knew shows the city as a square instead of the usual circle, perhaps an acknowledgement of topographical reality.[28] Matthew's copy may have been similar, or he may simply have adjusted the plan to reflect William of Tyre's description of Jerusalem as rectangular, a description that Matthew incorporated into his own *Chronica majora*.[29]

Outside Acre and Jerusalem it was a world map that supplied the cartography. All three versions have a straight coastline from Acre to Damietta – just as on the second Tournai map of Palestine and on Matthew's Oxford map – and the same seven places are shown on each. A little way inland all three show Arsuf (incorrectly as it lies on the coast) and B and D also show Gaza. All three mark the route from Jaffa to Jerusalem, passing through Ramla, which is shown on all three but named (*Rames*) only on D. Beyond Damietta the coastline abruptly turns eastward to give Palestine a southern coast; we need not assume that it appeared thus on the world map or that Matthew thought this was how it really was, for he may have simply wanted to include Alexandria and had run out of space to extend the coast further to the right. To make the point clear, both B and C note that Alexandria is on the sea (*ki set* [*sus*] *mer, sus mer*)[30] and on D wording along the aberrant coastline reads 'This is the way towards Alexandria' (*Ceste est la voie vers Alisandre*). All three versions show Cairo on one side of the Nile, and the new Babylon on the other. Maps B and C show Mansourah on what B names as a branch of the river, and on both B and D is a drawing of a crocodile.

22 Simek, 'Hierusalem', pp. 133–5, 137–40, 142, 145–7.

23 I.e. respectively the Dome of the Rock and the El Aqsa mosque.

24 On all three the Holy Sepulchre is shown as a near-circle with a projection, like a retort, perhaps deriving from the plans of Adomnan and Bede (above, p. 8).

25 In Anglo-Norman French, *sor* can mean not only *red* but also *golden* (*Anglo-Norman dictionary*, ed. by William Rothwell et al., Modern Humanities Research Association 8 (London, 1977–92), p. 723).

26 Revelation, 21.10–17

27 As Connolly, *Maps of Matthew Paris*, pp. 110, 114–15, assumes.

28 Heydenreich, 'Jerusalem-Plan', plate 1; Simek, 'Hierusalem', pp. 127–8, 136, 141. Ingrid Baumgärtner, 'Die Wahrnehmung Jerusalems auf mittelalterlichen Weltkarten', in *Jerusalem im Hoch- und Spätmittelalter*, ed. by Dieter Bauer et al. (Frankfurt, 2001), pp. 271–334, discusses the ways Jerusalem was shown on medieval world maps, with reference to the city plans.

29 *Matthaei Parisiensis chronica majora*, ed. Luard, II, p. 109; *Guillaume de Tyr*, ed. Huygens, I, p. 383.

30 B is slightly damaged at this point.

East of Jerusalem we have a clear demonstration that what mattered to Matthew was placing features in the right part of the map; their relative positions there were less important. On all three versions the rivers Jor and Dan combine to form the Jordan; there is no sea of Galilee but the river passes through the Dead Sea and a little beyond. On B and C the Jor and the Dan rise in the mountains of Lebanon, and also on B and C the river's lines are unbroken across the Dead Sea, as in the sea of Galilee on the Sawley world map and as on Matthew's Oxford map.[31] On all three Nazareth lies south of the Dead Sea, while Bethlehem, with star above, lies between Nazareth and Jerusalem. On B and C Mount Tabor comes between the Dead Sea and Nazareth,[32] on C and D Jericho appears north of Bethlehem and on D alone is the dwelling of Zacharias and Elizabeth, pictured as a range of hills.

North of the Jordan and its sources we move to the map's left-hand page, where the plan of Acre is the principal feature. East of Acre is a two-humped camel, loaded with bales, that may well have come from a world map: there is one unladen in Bactria on the Hereford map, another on the Ebstorf map at Jerusalem. On B and C it stands alone, but on D a man leads it; on C it is named, *camelus*, and nearby is another picture of a buffalo (*bubalus*), attached to what may be the swingletree of a plough, and a laden mule (*mulus*) led by a man partly concealed in the gutter between the two pages.[33] All three versions show Damascus with the rivers Abana and Pharpar; the road from Damascus to Acre, passing through Safed, appears on B and D but not on C. North of Damascus C shows Nineveh with a picture of Jonah being swallowed or ejected by the fish, confusingly labelled *coco*, i.e. *cocodrillus*, crocodile;[34] and C and D both show Noah's ark in the mountains of Armenia. At the north end of the map a quadrant in the top-left corner of all three versions marks the wall of the Caspian Mountains behind which Alexander the Great enclosed the forces of Gog and Magog. The home of the Old Man of the Mountain appears on all three maps, and so too does Antioch, along with the Black Mountain (identified as a monastery on B and D). The coast here is lost on D, but both B and C show Tyre as a near-island that juts out into the Mediterranean.

Besides the world map, Matthew must have drawn on other sources for a few of these cartographic features, perhaps again including visitors to St Albans. Daron appears on no other contemporary map, nor does Mansourah, which was in the news in 1250 as the site of the battle and of King Louis IX's captivity. But certainly a world map provided most of the material. Matthew himself drew a world map on a page now at the end of MS 26, but originally at the front with the other illustrations **[44]**.[35] In a note above it he describes it as abridged (*summatim*) from a world map by Master Robert *de Melckeleia*[36] and from one at Waltham Abbey, and it is indeed very much abridged: it extends no further east than the Caspian Sea and the Persian Gulf, there are no islands in the ocean, not even Britain, and few towns are named. In the same note he mentions another world map that he put in an ordinal, a service-book, that he had written; he had copied it from one in the king's room at Westminster.

We know of the Westminster map from other sources, but this note is our only evidence for Master Robert's and the Waltham Abbey maps. It adds to our understanding that detailed, encyclopaedic world maps, though not numerous, were a recognised feature of thirteenth-century English culture. Robert's name comes from

31 Above, pp. 69-70.

32 A note on B and D that seems to give the distance from Bethlehem to Nazareth as 3 leagues is probably misplaced, and is really the distance from Nazareth to Mount Tabor, as in *Itinéraires*, ed. Michelant and Raynaud, p. 101.

33 In *Matthaei Parisiensis historia anglorum*, ed. Madden, III, p. 119, Matthew tells how *bubali* of both sexes were sent to Richard, Earl of Cornwall, in 1252, so that they could be bred in England.

34 Presumably standing in for the crocodile that B and D show on the Nile.

35 Cambridge, Corpus Christi College, MS 26, f. vii v.

36 An added note beside Robert's name, *in ecclesia*, 'in the church', suggests that the map may have been a wall-painting. Though the words might be thought to apply to the adjacent note of King Alfred's ideal of spending one-third of one's time in divine office (*in diuinis obsequiis*), a pen-stroke seems to link it rather with Robert's name.

a Hertfordshire place-name that survives in the form Mentley.[37] He was rector of Clothall in Hertfordshire from 1237 until he died in 1271, but he probably led his life on a wider stage.[38] In 1261 he was described as a clerk of the king and we catch glimpses of him on royal business between 1255 and 1262.[39] He was presumably related to the older Master Gervase of Mentley, pointing to a cultured background:

37 J.E.B. Gover et al., *The place-names of Hertfordshire*, English Place-Name Society 15 (Cambridge, 1938), pp. 197-8.

38 Robert Clutterbuck, *History and antiquities of the county of Hertford* (3 vols, London, 1815-27), III, p. 504. That he died in, not before, 1271 when his successor was instituted, is shown by his acknowledgment of a debt in that year (*Close rolls 1268-72*, p. 416).

39 *Calendar of patent rolls 1247-58*, p. 508; *Close rolls (supplementary) 1244-66*, pp. 22, 28, 33; *Close rolls 1259-61*, pp. 269-70; *Calendar of liberate rolls, 1260-7*, pp. 64, 105.

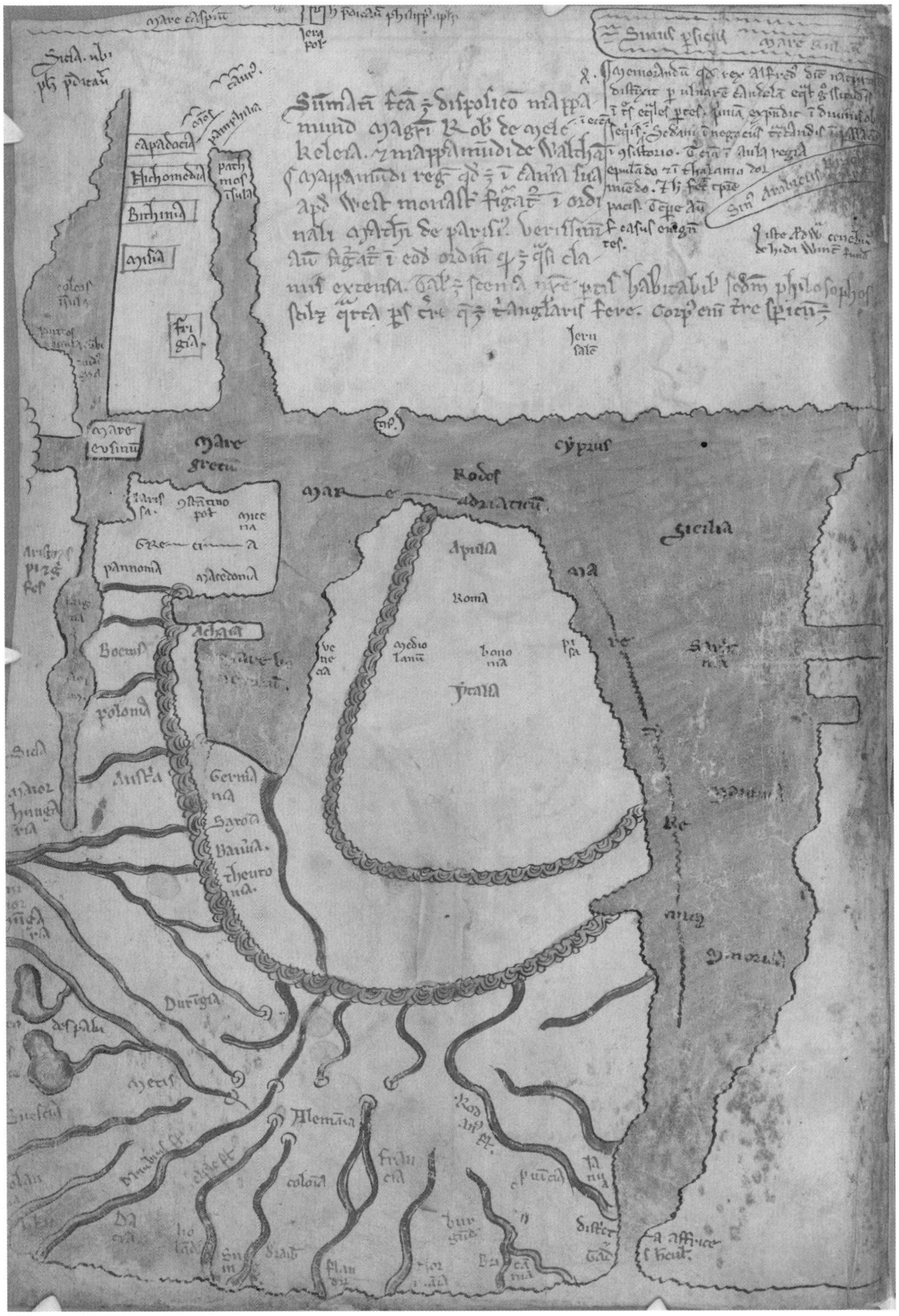

44. World map by Matthew Paris, mid-13th century; east at the top. Little more than an extract showing the Mediterranean and the Black Sea; Italy is the broad peninsula in the centre. Cambridge, Corpus Christi College, MS 26, f. vii v.

Gervase, who studied probably at Paris, perhaps at Oxford, wrote Latin verses and was described by Matthew as an outstanding astrologer (*optimus astrologus*).[40] Matthew may well have known other world maps too; what was probably the earliest of his four maps of Britain, in MS 14 C.vii, seems to have taken its outline from one world map, the other three from another.[41]

In their cartographic content, then, there is little to choose between the three maps. Each of the three contains at least one feature that appears on neither of the others, each of the three omits some feature that both others include. In their style of drawing, however, there are distinct differences between them. B is the most carefully drawn, the most minutely detailed of the three; C is rather less so and D, while still a neatly executed map, is somewhat sketchier than the other two. To take one tiny example among many, on B the two temples at Jerusalem are drawn, one with a dome, the other with a spire, with some architectural detail; on C each is simply a dome placed on a pediment, and on D one is a battlemented outline with a tower and the other is marked just by a small circle **[45]**.[42] But outside the plan of Acre it is not the

40 A.B. Emden, *Biographical register of the university of Oxford to A.D. 1500* (3 vols, Oxford, 1957–9), II, p. 1256; *Matthaei Parisiensis historia anglorum*, ed. Madden, II, p. 232; *Matthaei Parisiensis chronica majora*, ed. Luard, III, p. 43; IV, p. 493.

41 Harvey, 'Paris's maps of Britain', pp. 114–15.

42 This last is conceptually interesting, as it suggests that Matthew was thinking not pictorially but in terms of plan, as in the representation of the Holy Sepulchre (n.24 above).

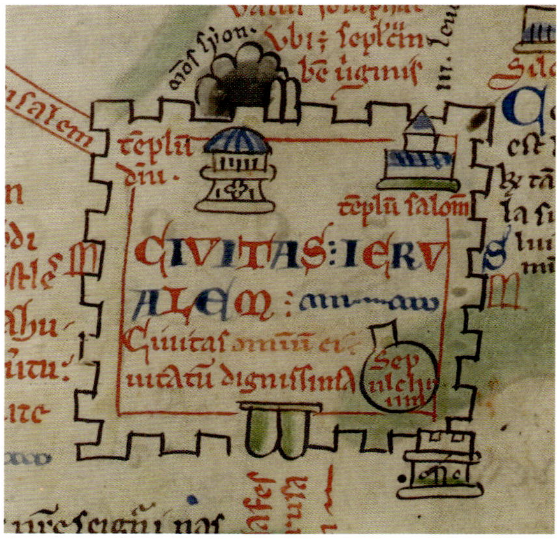

45. Jerusalem on the Acre map of Matthew Paris. Each of the two temples is drawn in some detail on map B (above) but simply as a dome on a pediment on C (top right), and on D (right) one is marked only by a small circle. BL, Royal MS 14 C.vii, f. 5r; Cambridge, Corpus Christi College, MSS 26, f. iv r; 16, f. iv r.

cartographic features that dominate all three maps, but the great blocks of text, some in red ink, some in black, with blue paragraph marks on B, red and blue initial letters on C and D. There is rather less text on B than on D, less on D than on C, where the extension to the left-hand page offered extra space, but, like the first Tournai map of Palestine, all three maps were meant to be read, not merely looked at. The blocks of text and other inscriptions are mostly in French, but some are in Latin, rather more on B than on C or D.

On all three maps the texts are broadly geographical: unsystematic notes on towns and regions, their wealth, their peoples, whether they were in the hands of the Saracens and so on.[43] On B and C in the bottom right-hand corner (lost from D) is a long note on Africa, 'This land that is on the right' (*Ceste terre ki est a destre*) – and thus beyond the edge of the map. Two rather different texts that appear on C and D but not on B probably come from an account of Jerusalem and the Holy Land that was attached to some copies of the so-called Rothelin continuation of William of Tyre's chronicle.[44] The story of the tree that bent down to give its fruit to the Virgin is there, and the account in C of the miraculous picture of the Virgin at Saidnaya (*Sardainne*) may well have had the same source; like Matthew, it mentions the picture's Greek origin and he need only have changed the dimensions from ells (one *aune* high, one-half *aune* broad) to the perhaps more familiar feet (three or four *pez* high and less broad). It does not, however, include the story of the chickpeas that Jesus turned into pebbles, another notable text that is on C and D but not on B,[45] and Matthew probably drew on a range of sources for all the texts on the maps. On B and D we are told that the route from Gaza to the new Babylon (beside Cairo) passes through *la berrie*, the desert.[46]

Konrad Miller sees links with the same account of the Holy Land that contributed to Matthew's Oxford map and points out that, while this looks very different from the Acre map, the two maps have interesting features in common. He mentions as examples the Oxford map's list of sultanates and its note of the creation of Adam;[47] we find both on C and the creation of Adam on D as well, and there are indeed other points of contact between the maps. They are the only known medieval maps – of the world or of the region – that give the distances in day's journeys between places in the Holy Land, the only ones that name Saidnaya or that spell Gaza with *r*: *Gazre*, *Gazeres*, *Gazres*. The Oxford map mentions the crocodiles in Egypt and B and D both picture one. None of this need point to direct borrowing. Indeed, the lists of sultanates differ, Adam's creation is entered on the borders on Egypt on the Oxford map but at Damascus on C and D, the distances in day's journeys are between different places on the two maps and on the Acre map the references to Saidnaya are linked to the miraculous picture of the Virgin Mary – the Oxford map enters the name (*Sardenay*) but without mentioning the picture.[48] All the same, the two maps have enough detail in common to suggest that Matthew at least took some ideas from the Oxford map when he came to compile the Acre map.

This assumes that the Oxford map is older than the Acre map, which is likely but by no means certain. All we know for certain is that the Oxford map cannot be earlier than 1246 and that of the three versions of the Acre map B and C cannot be earlier than 1252 for, as Vaughan points out, what they tell us about Armenia came

43 The legends on all three versions of the map are transcribed in *Itinéraires*, ed. Michelant and Raynaud, pp. 123-39, except that the editors understood that the whole of the right-hand page of D was missing, not just its lower part; thus the frequent note 'Manque à D' is often incorrect.

44 *Itinéraires*, ed. Michelant and Raynaud, pp. xxiv-xxvii, 141-75. While the text in this form dates from 1261, its basis was probably an earlier account.

45 On C a *lapsus calami* by Matthew (putting at one point *chiches* for *peres*) obscures the point of the story.

46 From the Arabic *barrīya*, an explanation that I owe to Prof. Denys Pringle. Konrad Miller links the word with an account of the routes to the new Babylon, known from a late-thirteenth-century version, that calls the staging-posts on this road *berith* (Miller, *Mappaemundi*, III, pp. 93-4; *Itinéraires*, ed. Michelant and Raynaud, p. 242).

47 Miller, *Mappaemundi*, III, p. 156.

48 Reinhold Röhricht, in his typeset reproduction of the Oxford map (Röhricht, 'Karten und Pläne VI', plate VI), reads *Simulacrum Sardenay*, but *Simulacrum* is in fact a misreading of *Similiter* from an adjacent note.

from visitors to St Albans in that year, as recorded in the *Chronica majora*.[49] Indeed, the itinerary in MS 14 C.vii (and thus map B) cannot be earlier than 1253, for it refers to the offer of the Sicilian crown to Richard, Earl of Cornwall, in that year, and is unlikely to be later than 1254 as it makes no mention of its acceptance then by Henry III on behalf of his young son Edmund.[50] It does not follow that, because C is in the first volume of the *Chronica majora*, D in the second and B in the third, the maps – and the rest of the introductory material – were produced in this order, and there is some reason for supposing that B, in MS 14 C.vii, is the oldest of the three.[51] It was more carefully executed than the other two, its outline of the walls of Acre is more accurate and its inscriptions are shorter with a higher proportion of Latin.[52] On the back of its right-hand page is what is probably the earliest of Matthew's four maps of Britain; what is probably a later map of Britain is on the back of D.[53] The complex thus created of itinerary and Acre map will then have served as a prototype for MS 16 (with map D) and MS 26 (with map C). We may reasonably guess that D, with rather more text than B, was drawn next and that C, with the greatest amount of text and the extended left-hand page, was the last of the three to be drawn – but this can be only a guess. This might mean that the itinerary, the Acre map and the rest of the preliminary material were first put together to accompany the *Historia anglorum* in MS 14 C.vii and were then replicated, with variants, to introduce the first two volumes of the *Chronica majora* in MSS 26 and 16 – but this too is mere speculation.[54]

A copy of the preliminary material in MS 26 was made in the thirteenth century. It is not in Matthew's hand but was presumably made at St Albans, for it survives bound into a register that contains fourteenth- and fifteenth-century copies of charters and estate records of the abbey.[55] This register was among the manuscripts of Sir Robert Cotton (died 1631) and, like many others in his collection, was badly damaged in the fire at Ashburnham House, Westminster, in 1731. The map of Palestine is partly burnt away, and the parchment has shrunk severely through the heat. However, enough is visible to show that it is a close copy of C **[46]**. Arsuf and the route from Jaffa to Jerusalem seem to have been omitted, perhaps through lack of space, but otherwise there are only minor differences in spelling and drawing: *Sepulcrum* for *Sepulchrum* at Jerusalem, for instance, and the planks of the boats are drawn with vertical as well as horizontal lines.

Happily the three manuscripts of the *Chronica majora* (with the *Historia anglorum*) met with better luck than this register, though they had varying fortunes. Before they passed into the library of Archbishop Matthew Parker and thence, on his death in 1555, to Corpus Christi College at Cambridge, MS 26 had belonged to Edward Aglionby of the Warwickshire gentry and MS 16 to the antiquary Robert Talbot (died 1558) and to Sir Henry Sidney, who was three times Lord Deputy of Ireland.[56] Both these volumes presumably remained at St Albans until the abbey's dissolution in 1539, but MS 14 C.vii belonged successively to Humfrey, Duke of Gloucester (died 1447), and John Russell, bishop of Lincoln, on whose death in 1488 it may have returned to St Albans. After 1539 it was for a time in the royal library, but then belonged first to Henry Fitzalan, Earl of Arundel (died 1580), then to John, Lord Lumley (died 1609); his library was acquired by Henry, Prince of Wales, who died in 1612, and MS 14 C.vii passed, with the rest of the royal manuscripts, to the British Museum on its foundation in 1753.[57]

49 Vaughan, *Matthew Paris*, pp. 239, 245.

50 As first pointed out by George B. Parkes, *The English traveller in Italy*, I: *The Middle Ages (to 1525)* (Rome, 1954), p. 180n.

51 Both Comte Riant in *Itinéraires*, ed. Michelant and Raynaud, pp. xxiii-xxiv, and Miller, *Mappaemundi*, III, pp. 90-1, speak of B as version 1 of the map, C and D as version 2, but without specific commitment to this chronological order.

52 The late addition of battlements to Montmusard on B (above, p. 82) may mean that the informant was still at hand when the plan in B (presumably a fair copy from a first draft) was drawn.

53 Harvey, 'Paris's maps of Britain', pp. 114-16.

54 I pass over two theories concerning MS 14 C.vii that appear to owe nothing to the actual evidence. According to Lewis, *Art of Matthew Paris*, pp. 326, 332, the itinerary in MS 14 C.vii was originally drawn on one side only of four sheets of parchment aligned side by side; they were then pasted back to back to form two leaves of the book – with the addition of the route through Italy on the back of the left-hand page of map B. This would be of great interest if it were true, but unfortunately it is not; the first four pages of the itinerary were executed on both sides of two pieces of parchment that, though now kept separately, must from the start have been two leaves of the quire that came to be included in this volume. Connolly, *Maps of Matthew Paris*, pp. 173-91, suggests that the *Historia anglorum* in MS 14 C.vii, with the introductory material including map B, was executed after Matthew's death as a de luxe manuscript for presentation to the king. For such a suggestion to be taken seriously it would need to be backed up by rigorous and detailed analysis of the hand; I certainly see no reason to doubt that both the preliminary material and the text are in the hand of Matthew.

55 BL Cotton MS Tiberius E.vi, ff. 3v-4r; G.R.C. Davis, *Medieval cartularies of Great Britain and Ireland*, ed. by Claire Breay et al. (London, 2010), p. 169.

56 James, *Manuscripts in Corpus Christi College, Cambridge*, I, pp. 50, 55.

57 Warner and Gilson, *British Museum: catalogue of Royal manuscripts*, I, pp. xviii-xix; II, pp. 135-6.

Matthew's Acre map has long been better known to antiquaries and historians than any of the other maps of Palestine discussed here – a little perversely, given that it is arguably the one that offers the most limited view of the Holy Land. A manuscript copy of B, with the itinerary, is identified by Sir Frederic Madden as the work of William Camden (1551-1623), antiquary and historian **[47]**,[58] and in 1780 Richard Gough described B in print, with an engraving of the itinerary's first section.[59] In 1844 the British Museum's catalogue of manuscript maps described B, with the itinerary, as 'A map of the stations for a pilgrimage from England to the Holy Land, consisting of rude drawings of the towns from London to Jerusalem, with the names in French' and, probably soon after, lithographic reproductions were published in Paris in a collection edited by E.-F. Jomard.[60] In 1882, following the introduction by Count Riant which described these reproductions as a 'soi-disant facsimile' that rendered the map's inscriptions 'absolument illisibles', Henri Michelant and Gaston Raynaud included the texts from all three versions of the map, as well as from the early-modern copy, in their edition of medieval accounts of the

58 BL Lansdowne MS 253, ff. 194v, 197r. The note by Madden is on f. 197v.

59 Richard Gough, *British topography* (2 vols, London, 1780), I, p. 85, plate VII.

60 *Catalogue of manuscript maps, British Museum*, I, p. 14; E.-F. Jomard, *Les monuments de la géographie* (Paris, [1842-62]), nos VIII, IX.

46. Thirteenth-century copy of map C, badly damaged, like many others of the Cotton manuscripts, in the fire at Ashburnham House in 1731. BL, Cotton MS Tiberius E.vi, ff. 3v-4r.

Holy Land written in French.[61] In view of this, when Konrad Miller described the maps in 1895 he gave the inscriptions only from B and for the longer texts gave only the opening words in place of his usual full transcription.[62]

In his article of 1953 and monograph of 1958 Richard Vaughan effectively settled the question of Matthew's handwriting that had divided scholars since the 1870s. On the basis of his careful analysis there can be no reasonable doubt that all three versions of Matthew's Acre map – as well as his Oxford map – are in Matthew's own hand.[63] Vaughan's account of the Acre map and the itinerary treats them as separate productions; in this he follows Miller and the geographer C.R. Beazley, who in 1901 suggested that they have been linked together only by their chance juxtaposition in the manuscripts: 'Matthew's so-called *Itinerary from London to the Holy Land*... is not really a connected whole; it is the result of combining two different works.'[64] However, given that in each manuscript the itinerary immediately precedes the Acre map, while the other preliminary material is in no particular order, given that the itinerary, once it reaches south Italy, becomes ever less route-like, ever more map-like, given too the parchment extensions to the itinerary's last page, which even contribute to map C, it

61 *Itinéraires*, ed. Michelant and Raynaud, pp. xxii-xxiv, 123-39 (but cf. n.43 above).

62 Miller, *Mappaemundi*, III, pp. 90-3.

63 Vaughan, 'Handwriting of Matthew Paris'; Vaughan, *Matthew Paris*, pp. 35-48, 245n.

64 Vaughan, *Matthew Paris*, pp. 241-2, 244-5, 247-50; Miller, *Mappaemundi*, III, pp. 84-5; Beazley, *Dawn*, II, p. 588.

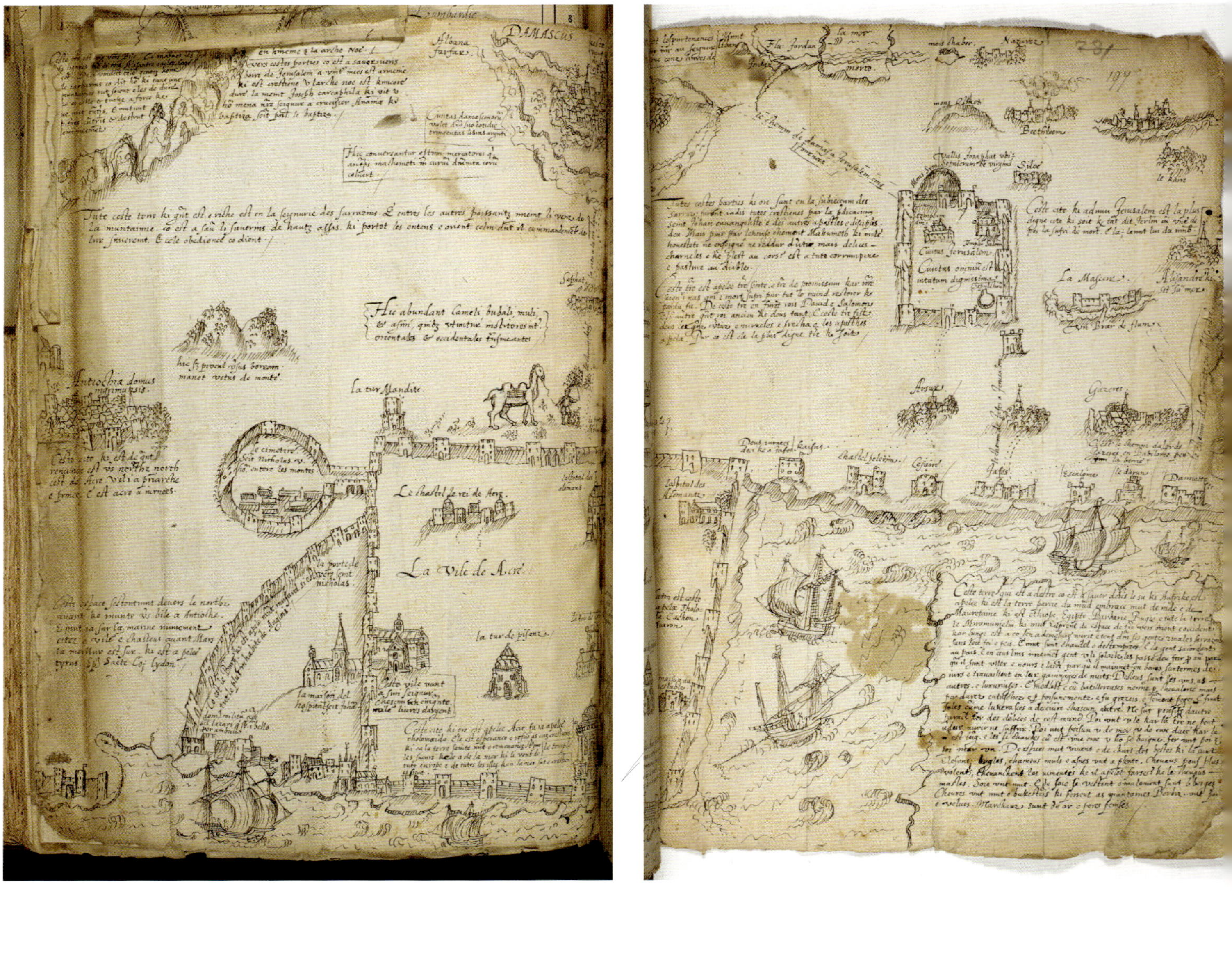

47. Copy of map B by William Camden, historian and antiquary (1551-1623). BL, Lansdowne MS 253, ff. 194v, 197r.

seems most likely that Matthew himself saw the Acre map as the fitting end of the itinerary.

This is certainly how it appears to two art historians who examine both together as forming effectively a virtual pilgrimage to the Holy Land: Suzanne Lewis in her book of 1987 and Daniel K. Connolly in his article of 1999 and book of 2009.[65] Both pay especial attention to the strip-maps of the itinerary, and Connolly also attaches much conceptual importance to the parchment extensions to its last page – all of which is fully understandable, for these are the most startlingly original and imaginative parts of the whole. But they must not be allowed to eclipse the importance of the Acre map of Palestine; it too is an original and imaginative production.

65 Lewis, *Art of Matthew Paris*, pp. 323-64; Connolly, 'Imagined pilgrimage'; Connolly, *Maps of Matthew Paris*. Johannes Weiss, 'Ein dynastisch-territoriales Bild ihrer Zeit? Die Itinerar- und Palästinakarten von Matthaeus Parisiensis', *Mitteilungen des Instituts für Österreichische Geschichtsforschung*, 116 (2008), pp. 250, 252-3, is also in no doubt that map and intinerary form a single whole, and this interpretation underlies the whole of Sansone, *Mattheus Paris*.

9 THE EARLIEST LARGE BURCHARD MAP
circa 1300

Florence, Archivio di Stato, Carte nautiche, geografiche e topografiche 4 **[48]**. Parchment, 51.5 × 168 cm; four membranes, glued together with 15-20 mm overlap, width of each (left to right) 42.5 cm, 43 cm, 44.5 cm, 43 cm. Brown and red ink, with beige, mauve, dull olive, pale olive and red paint. Dorse blank except for the stamp of the Archivio di Stato and modern pencilled number.[1]

THIS IS THE EARLIEST KNOWN MAP to be based on Burchard of Mount Sion's description of Palestine, but it is by no means the last; every later map discussed here owes a great deal to Burchard's work. It is a remarkable map, and Burchard's is a remarkable book, in a class by itself among the many medieval accounts of the Holy Land. All else apart, it is long and detailed; with the editor's notes the Latin text extends to some seventy-five printed pages.[2] But more important than this, it is written with much local and scriptural knowledge, with verve and with considerable charm. Some account of Burchard's work, of its different versions and editions, is essential before looking at the maps to which it contributed so much.

Of Burchard himself we know only what we can glean from his book – mostly, indeed, from a shorter version that he addressed to another Burchard, a Dominican friar at Magdeburg.[3] Here he refers to himself too as *frater*, a friar, and he too was a Dominican if we are to believe the heading of his work in some manuscripts.[4] Like his friend, he was a German speaker, for he explains that a point on the road from Jerusalem to Jericho where there were many robbers was given a name that in German (*theotonice*) would be *Rotenburch*, from the red blood that was shed there.[5] Like his friend, he knew Magdeburg, comparing its size with that of Jerusalem and noting that the distance from Acre to Nazareth was about the same as the 25 kilometres from Magdeburg to Barby – though his memory was at fault, for it is nearly half as far again. J.C.M. Laurent, who edited Burchard's work, points out that Burkhardt was a common name in the family of the counts of Barby, to which he perhaps belonged; 'of Mount Sion' would be his name as a friar.[6]

Burchard spent at least two years in the Levant, possibly much longer. He could read Arabic, for he described the beliefs of the Saracens from having read the Koran.[7] Certainly he had been in the Holy Land long enough to know it well and his account was based on his own experience: 'I would wish the reader to know that I have included nothing in this description except what I saw with my own eyes when I was in the place itself or, when I was unable to gain access, what I saw standing on some mountain or in another suitable place; and I have noted down what I have learnt from Syrians, Saracens or other inhabitants of the land, diligently questioning them.'[8] While he specifically restricted his work to the Holy Land itself, he had travelled in

1 I am most grateful to Dr Marina Laguzzi of the Archivio di Stato in Florence for her kindness and help when I worked on the map in 2001; she enabled me to inspect the original map as well as the excellent full-size photograph that is normally made available. The drafting lines and grid are mostly visible only on the original map.

2 *Peregrinatores*, ed. Laurent, pp. 19-94.

3 This and the following paragraph draw heavily on *Peregrinatores*, ed. Laurent, pp. 3-4, and on Pringle, *Pilgrimage*, pp. 46-51. I am most grateful to Prof. Pringle for allowing me to see his work before publication and to quote from his translation.

4 Thus both in Florence, Biblioteca Medicea Laurenziana, ms Plut. 76.56, f. 94r, and in Hamburg, Staats- und Universitätsbibliothek, Codex geographicus 59 (*Peregrinatores*, ed. Laurent, p.6), he is described as *ordinis predicatorum*.

5 *Burchardus, Liber de descriptione*, p. 108.

6 *Peregrinatores*, ed. Laurent, p. 4; *Grosses vollständiges Universal-Lexicon* (64 vols, Halle and Leipzig, J.H. Zedler, 1732-54), III, cols 425-6.

7 *Peregrinatores*, ed. Laurent, p. 53.

8 Pringle, *Pilgrimage*, p. 242.

the lands beyond it, giving the itinerary of a voyage he had made from Cappadocia and Cyprus to Egypt, and recounting a conversation he had had with the sultan of Egypt about the cultivation of balsam.[9] On the other hand he is careful to explain that he had never been in the areas east of Galilee.[10] Laurent showed from internal evidence that Burchard wrote after 1271 but before 1285, and Denys Pringle has narrowed this further to between July 1274 and May 1285. In some early printed editions of the text Burchard tells us that he was on Mount Gilboa on All Saints Day (1 November) in 1283 and, as we shall see, Laurent is not necessarily correct in seeing the year as corroborative – but invented – detail added by an editor.[11]

Burchard's account follows a systematic plan. He divides the country into seven sectors, set out fanwise from Acre, and after an introductory prologue he covers each in turn, moving round from north to south. The directions of the first six sectors, he explains, correspond to the six winds, that is, the six winds of the eastern half on Pliny's system, familiar to learned persons in the thirteenth century.[12] Some manu-scripts of Burchard's work have a diagram showing the sectors, with the names of the winds **[49, 50]**.[13] In the book each sector is given a new heading, though in the text printed by Laurent they are named rather confusingly: they start with the first, second, third and fourth divisions (*divisio*), but the fifth and sixth sectors are called the second and third divisions of the eastern quarter (*divisio quarte orientalis*) and there is then a seventh sector, the first division of the southern quarter (*divisio quarte australis*), for places south of Acre along the coast. There was, of course, no great precision; the sectors were the routes followed by a traveller, not lines drawn on a map. All the same, the scheme would have worked better had the centre of the fan not been so far north, for most of the area described lay south of Acre; the sixth sector in particular is disproportionately large, taking in Jerusalem and most places further south. However, it is interesting that Burchard should have thought of the country in this – to us – semi-cartographic way.[14] Following the seventh sector are substantial accounts of the Holy Land's vegetation and wild life, of its people's religions and of Egypt.[15]

Burchard wrote, he explains, for those Christians who, unable to make the journey to the Holy Land, nonetheless eagerly wished to envisage the places where the events of the Bible took place, the places where Christ lived and died. Throughout, he describes what the visitor, the pilgrim, would find, so that his readers could imagine themselves there, sharing in what Burchard saw and experienced in his travels. He gives many measurements. Most distances are in leagues (*leuce*) but the longer ones are defined as so many day's journeys, shorter ones in terms of a bowshot or a stone's throw.[16] Thus two moderate leagues (*modice leuce*) separate Mount Hermon from Mount Gilboa; the Dead Sea is six leagues wide, five day's journeys long; Bethlehem lies two leagues from Jerusalem, a bowshot off the road to Hebron, and within the town the church of Sts Paula and Eustochium is a stone's throw from the church of the Nativity.[17] He provides detailed dimensions of some holy sites: at Jerusalem the cave of the Holy Sepulchre is eight feet wide and eight feet long, the site of the Crucifixion lies 120 feet away and is reached by going up eighteen steps.[18]

Burchard describes the monuments marking the holy sites, but he describes much else besides. Many towns and villages are merely mentioned by name, but of many others we are told a good deal more, of their present state or of the biblical events

9 *Peregrinatores*, ed. Laurent, pp. 31, 61, 93-4.

10 Ibid., p. 41.

11 Ibid., p. 52n.

12 Westrem, *Hereford map*, p. 12.

13 BL Add. MS 18929, f. 51r; Hamburg, Staats- und Universitätsbibliothek, Codex geographicus 59, p. 13. Another is in Munich, Bayerische Staatsbibliothek, Clm 569, f. 186v (Baumgärtner, 'Das Heilige Land').

14 This is discussed by Gautier Dalché, 'Cartes de Terre Sainte', pp. 586-92.

15 The account of Egypt, omitted from the manuscripts used for the text in *Peregrinatores*, ed. Laurent, is printed in *Thesaurus monumentorum*, ed. Basnage, IV, pp. 25-8.

16 Where he uses some other unit, it has probably been taken from another writer's account. Thus he specifically cites Josephus in giving the circuit of the walls of Jerusalem (excluding Mount Sion) as 33 *stadia* (*Peregrinatores*, ed. Laurent, p. 72).

17 *Peregrinatores*, ed. Laurent, pp. 51, 59, 78, 79. All these measurements are particularly liable to error in copying and there is clearly much variation between one manuscript and another.

18 Ibid., pp. 70, 71.

enacted there. Tripoli, nearly surrounded by the sea, has a large population of Greeks, Latins, Maronites, Nestorians and many others; cloth of silk and of camel's hair is woven there and the vines, olives, figs and other fruits of the land around are said to bring its lords 300,000 gold besants each year.[19] At Endor we are simply reminded that this was the home of the woman who conjured up the spirit of Samuel before Saul.[20] Some towns, once important, had declined or were deserted; Burchard describes the ruins of great buildings at Sebaste and Nablus, and tells how at Jericho there were now scarcely eight houses.[21] He had an eye for fortifications; he describes the town walls at Acre and mentions many castles, among them the Templars' castle at Safed, 'the strongest and most beautiful of all castles in my opinion', which had been disgracefully betrayed and lost to the Christians.[22] This had been in 1266 and is one of Burchard's very few references to recent events; he tells how the sultan had established his treasury at Kerak,[23] but – interestingly and significantly – scarcely mentions the current political situation, who was in control of one or other region or town. He tells us much more about land use and wild life: at Tiberias are large palm-trees with vines and olives in very fertile land, on Mount Tabor lions roam among the ruined buildings, at Caesarea Palestina he was threatened by crocodiles.[24] He appreciated landscape and liked panoramas; at Jezreel was a fine view over the whole of Galilee and beyond the Jordan, while from the hill at *Rama*, between Bethlehem

19 Ibid., p. 28.
20 I Samuel 28.7–25; *Peregrinatores*, ed. Laurent, p. 48.
21 Ibid., pp. 53–4, 55, 58.
22 Ibid., pp. 23, 34; Pringle, *Pilgrimage*, p. 257.
23 *Peregrinatores*, ed. Laurent, pp. 58–9.
24 Ibid., pp. 45, 47, 83.

49. Diagram of Burchard's description of the Holy Land, 14th century; east at the top. The six segments fanning out from Acre are the successive itineraries of Burchard's text; the lines bear the names of cardinal points or winds and in the segments are listed the places described. BL, Add. MS 18929, f. 51r.

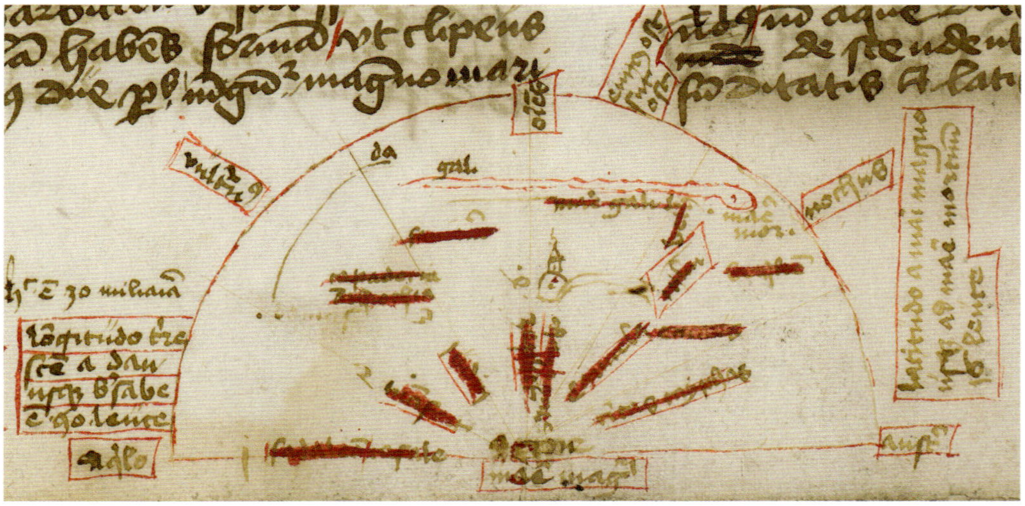

50. Diagram of Burchard's description of the Holy Land, 16th century; east at the top. This roughly sketched plan of Burchard's itineraries includes cartographic elements: the Jordan marked as a double line naming the sea of Galilee and the Dead Sea, and Jerusalem in the form of a circle and tower curiously reminiscent of the Tournai maps. Hamburg, Staats- und Universitäts-bibliothek, Codex geographicus 59, p. 13.

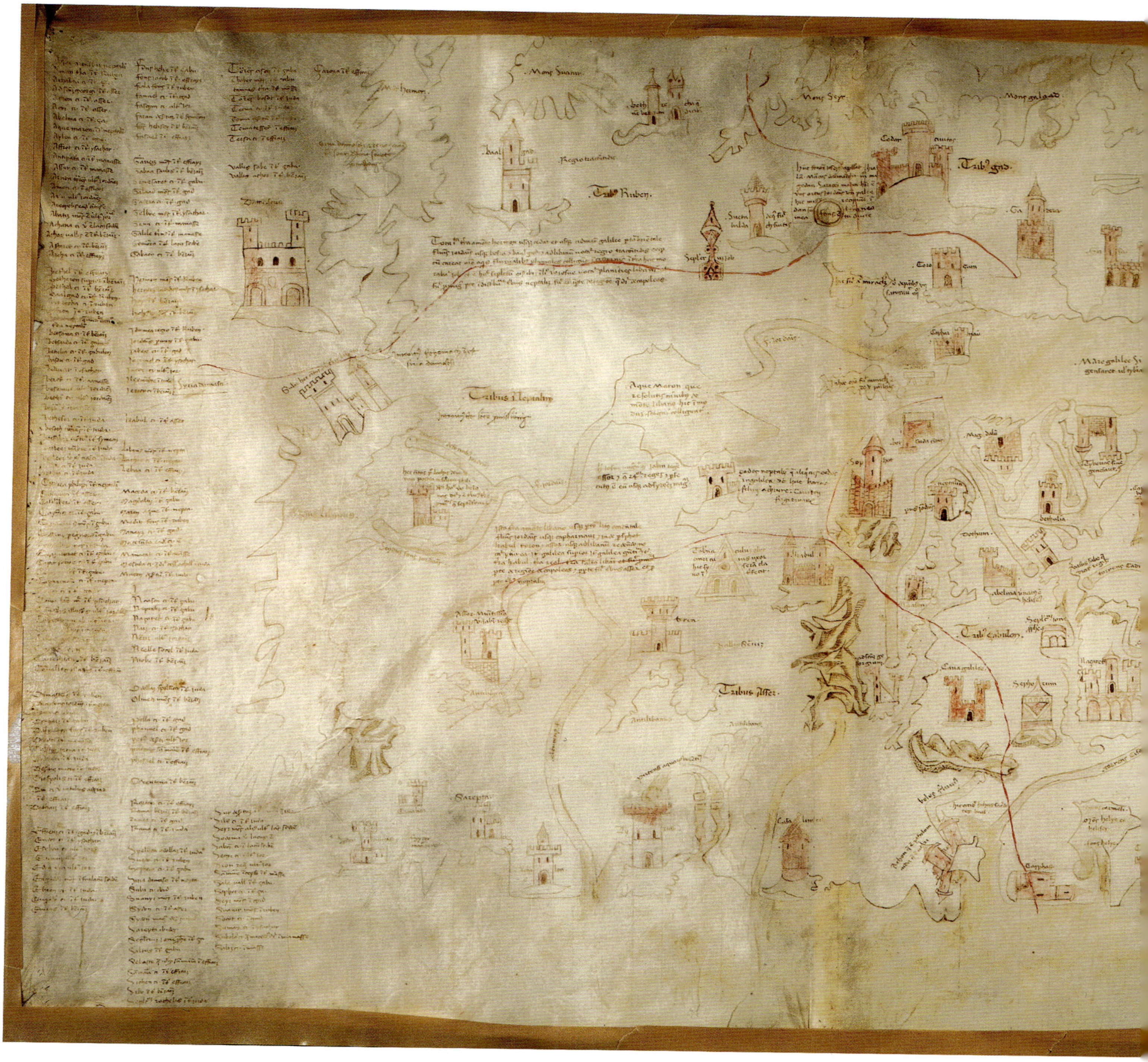

48. The earliest large Burchard map, circa 1300; east at the top. Our earliest, badly faded, example of this most impressive of all medieval maps of the Holy Land – it is over 5 feet wide. However, this copy derives, with demonstrable inaccuracies, from an earlier exemplar now lost. Florence, Archivio di Stato, Carte nautiche, geografiche e topografiche 4.

and Hebron, he saw the Jordan in one direction and in others Gaza, Beersheba and the whole coast from Jaffa to Ascalon.[25]

Much of the charm of Burchard's work lies in its accounts of his own experiences. Near the Dead Sea he wished to see the pillar of salt that had been Lot's wife, but he was told – incorrectly, he later learned – that the wild animals, the snakes and, above all, the Bedouins there made it too dangerous.[26] At Bethlehem he spent a whole night in the church of the Nativity, kissing alternately the spot where Christ was born and the manger where he was laid.[27] However, he was by no means a credulous traveller. Mount Gilboa was cursed by David after Saul and Jonathan were killed there: 'Ye mountains of Gilboa, let there be no dew, neither let there be rain upon you.'[28] But, Burchard says, this simply did not happen, for when he was on Gilboa it rained so hard that he was soaked to the skin.[29] At Hebron he saw the red earth from which God fashioned Adam; though quantities of it were taken away by pilgrims and by traders, who loaded it on camels for sale in Egypt, Ethiopia and India, the pit was always miraculously replenished. Burchard was clearly sceptical, for he notes that he forgot to ask if this was really true.[30] He helps the reader by giving the superseded or other alternative name for certain towns and explains how mountains with different names might all be part of a single range, as in the Alps.[31] At Cana, Burchard noted that here and nearly everywhere else the holy sites are below ground level, which he could only suppose was because the first churches built over them were demolished and replaced with new ones built on top of the ruins.[32]

Burchard's work was deservedly popular. More than a hundred manuscript copies are known and it was first printed in 1475.[33] By 1624 fifteen printed editions had appeared, besides translations into German and French.[34] However, there is no edition later than that of Laurent, who published it, together with three other accounts of the Holy Land, at Leipzig in 1864; a further account was added to a reprint in 1873. Laurent accompanied the text with an introduction in Latin and with many notes that identify the places named, give references to the Bible and other sources and comment on particular points; it is a tribute to him that in the course of a century and a half his work has not been superseded. He usefully analyses the early printed editions, showing that one group provides a text rewritten in the Latin of the sixteenth century instead of the thirteenth, while another offers no more than a wholly distorted abridgement.[35] More important, he shows that Burchard wrote two distinct versions of his work; besides the full version that Laurent printed there was the other shorter version, about a quarter the length of the full text, that Burchard addressed to his namesake at Magdeburg.[36] They are to be seen as separate works, not as variants of a single text.

This short version of the text was published by Heinrich Canisius in 1604 and reissued by Jacques Basnage in 1725.[37] A collation of this with four manuscripts was printed, but not published, by the Société de l'Orient latin in 1880, and this is our best edition of the short version.[38] As we have seen, it tells us more about Burchard himself than the full version, and at several points he adds a personal touch that is not in the full text; thus, nowhere had pleased him so much as the land around Sidon, while at Jerusalem one would have no trouble finding somewhere to live as so many houses were empty.[39] Laurent saw the short version sent to Magdeburg as a first,

25 Ibid., pp. 51, 80–1.
26 Ibid., p. 59.
27 Ibid., pp. 78–9.
28 2 Samuel 1.21.
29 *Peregrinatores*, ed. Laurent, p. 52.
30 Ibid., p. 81.
31 Ibid., p. 33.
32 Ibid., p. 44.
33 Thomas Kaepelli, *Scriptores ordinis praedicatorum medii aevi* (4 vols, Rome, 1970–93), I, pp. 257–9; IV, pp. 57–8; *Rudimentum noviciorum* (Lübeck, 1475), ff. 164r–188r.
34 *Peregrinatores*, ed. Laurent, pp. 11–17.
35 I.e. of the full version of Burchard's text (see below); *Peregrinatores*, ed. Laurent, pp. 12, 17, 18.
36 Ibid., pp. 10–11.
37 *Antiquae lectionis tomus I [–VI]*, ed. Heinrich Canisius (6 vols, Ingolstadt, 1601–4), VI, pp. 295–322; *Thesaurus monumentorum*, ed. Basnage, IV, pp. 1–28.
38 *Burchardus, Liber de descriptione*, pp. 92–117. This was intended as the first step in preparing a new edition, never achieved, of both the full and the short versions; the texts were printed with a wide margin so that scholars could enter variants from the manuscripts they had seen. The collation of the text of the short version was by W.A. Neumann.
39 *Burchardus, Liber de descriptione*, pp. 96, 109.

preliminary, text; Burchard wrote the full text later, after returning from the Holy Land.[40] However, in reading the two side by side it is difficult to avoid the impression that the short version is an abridgement of the full text that was already written, with small alterations or additions here and there: its phrasing, what it includes and what it omits, all point this way.[41]

This is of some importance when we look at the two maps directly based on Burchard's text – on the full version in each case. Laurent tells how, in the short version's letter to his friend, Burchard refers to a map: 'so that all these things can be the better pictured, I am sending you along with this a sheet of parchment on which they are all set out visually'.[42] If the short version mentioning the map was written before the full text, this map could not have been a copy of either the large Burchard map or the smaller one discussed in chapter 11, p. 141. However, if the short version was in fact abridged from the full text, it could well have been either of these maps – a point we shall return to.[43]

Laurent writes as if the reference to the map occurs in all copies of the short version of Burchard's text, but it is in neither Canisius's edition nor the text of 1880,[44] and it is fact to be found in one – and only one – of the three manuscripts of the short version that he saw in the university library at Wrozław (Breslau),[45] as well as in a manuscript at Munich.[46] Elsewhere too Laurent assumes that all copies of the short text must be identical: he specifically cites the Canisius edition as his source for quotations about Magdeburg that refer to its distance from Barby and the size of Jerusalem, but they differ from what Canisius prints and they come from the same Wrozław manuscript.[47] In editing the long version Laurent makes a few textual emendations,[48] but does not explore, or even mention the possibility, of differing manuscript traditions. It is on this basis that he takes the 1283 date, mentioned in the 1519 Venice printed text and some other editions, to be an editorial insertion;[49] *prima facie* it seems more likely that it is to be found in manuscripts that Laurent did not see. Certainly Laurent's edition is excellent as far as it goes. However, presented, as it is, with the authority and learning of nineteenth-century German textual scholarship, it is easy to overlook the fact that as a definitive edition of Burchard's work it leaves a great deal to be desired; indeed, Laurent cannot have expected it to be viewed in this light. The only manuscripts of Burchard's full text that he himself saw were two at Hamburg, one of the late fourteenth or early fifteenth century, the other of the sixteenth. He based his edition on these, noting their variants along with those in a manuscript at Berne lacking the first third of the text, of which Titus Tobler had sent him a transcript; in an appendix he listed further variants that Tobler had sent him from a manuscript at Basel.[50] Clearly these were simply manuscripts that lay conveniently to hand;[51] there

40 *Peregrinatores*, ed. Laurent, p. 10.

41 That the full text notes how the Saracens had now taken away the stone near Capernaum bearing Christ's imprint, while the short text mentions the stone as still there, might point to the opposite conclusion; but other explanations are possible (*Peregrinatores*, ed. Laurent, pp. 35-6; *Burchardus, Liber de descriptione*, p. 101).

42 'Que omnia, ut melius possint ymaginari, mitto vobis simul pellem, in qua omnia ad oculum figurantur' (*Peregrinatores*, ed. Laurent, p. 10).

43 In one 15th-century copy the full version of the text is entitled 'Treatise on the map of the Holy Land: how it may be painted' (*Tractatus de mappa Terre Sancte quomodo possit depingi*), presumably simply a tribute to the work's systematic plan and topographical detail (Gautier Dalché, 'Cartes, réflexion stratégique et projets', p. 81).

44 Nor in the copy of the short version in BL Add. MS 15835, ff. 163r-195v.

45 *Peregrinatores*, ed. Laurent, pp. 6-7, 10-11; Wrozław, Biblioteka Uniwersytecka, MS I F 221, f. 232v. I am grateful to Dr Jan Gromadzki for arranging for me to have microfilms of the three Wrozław manuscripts. In *Peregrinatores*, p. 10, Laurent also quotes Burchard in the letter to his friend as referring to an 'added geographic table' (*addita tabula geographica*), but the phrase does not occur there in any of the three manuscripts. Laurent's introduction and notes are in Latin, and it is not impossible that he accidentally attributed to Burchard a phrase that really belongs to his own text; its wording is a little unlikely for the thirteenth century.

46 Munich, Bayerische Staatsbibliothek, Clm 569, ff. 184r-210v (Baumgärtner, 'Das Heilige Land').

47 *Peregrinatores*, ed. Laurent, pp. 3-4; *Thesaurus monumentorum*, ed. Basnage, IV, pp. 14, 17; MS I F 221, ff. 235v, 239v. In the other Wrozław manuscripts that Laurent saw, MSS I F 277 and IV F 199, these passages are effectively the same as in the text printed by Canisius.

48 E.g. *Peregrinatores*, ed. Laurent, p. 32, n.137; p. 35, n.163.

49 Ibid., p. 52n.

50 Ibid., pp. 5-7, 100. He notes – not from vanity, he explains, but from the pleasure of having their support – that the Basel readings confirm some of his emendations.

51 The two Hamburg manuscripts, now in the Staats- und Universitätsbibliothek, were then in the Stadtbibliothek, where Laurent (1810-76) was a *Registrator* from 1835 to 1851, then Secretary until 1862, when he moved to Neuendettelsau in Bavaria to study theology (Hans Schröder et al., *Lexikon der hamburgischen Schriftsteller bis zur Gegenwart* (8 vols, Hamburg), IV, fasc. 2 (1863), pp. 384-6; Karl Bader, *Lexikon deutscher Bibliothekare*, Zentralblatt für Bibliothekswesen, Beiheft 55 (Leipzig, 1925), p. 144; these two works, with Friedrich August Eckstein, *Nomenclator philologorum* (Leipzig, 1871), p. 323, give Laurent's second name variously as Carl, Karl or Christian, and his third as Mauritz or Moritz).

is no reason to suppose they are the best, or even particularly good, copies of what Burchard originally wrote.[52] When we compare with Laurent's text the large and small Burchard maps, we are comparing what may well be an imperfect copy of Burchard's work with what are certainly, as we shall see, imperfect copies of the maps.

The earliest large Burchard map is in the state archives of Florence, but how it got there or where it came from we do not know. At the right-hand end is a stamp with crown and the initials 'R A D', standing for 'Regio Archivio Diplomatico'; this was applied to material gathered into this archival series after 1820.[53] The first, brief, notice of the map appeared in 1878, in a published inventory of oriental records and of maps in the state archives of Florence and Pisa; it was there assigned to the sixteenth century.[54] In 1891 Reinhold Röhricht published a redrawn and typeset reproduction, with a short discussion, in the first of his series of articles on early Palestine maps and plans;[55] the few of his readings that can be questioned are listed in Appendix 1 on pp. 105-6. Having consulted other authorities, he dated the map to the end of the thirteenth century and argued that it was drawn in Italy, as the letter *Z* is written in the form of *C* with a cedilla; there seems no reason to question either conclusion.[56] Röhricht looked at the possibility of a connection with the grid map that illustrated the work of Marino Sanudo (chapter 10, pp. 107-9) and two years later this was discussed in a much longer article by Cornelio Desimoni.[57] Both authors – almost certainly correctly – saw the present map as a little older than the grid map, and they were thinking in terms of a common original.

Since 1893 the only published notice of the map is a short entry in an exhibition catalogue of 1981.[58] This seems quite extraordinary, for it is one of the outstanding monuments of medieval cartography, but one reason for this neglect may be that it is badly faded. However, its sheer size is impressive: drawn on four sheets of parchment, it is 5½ feet long and 20 inches deep. East is at the top and the more or less straight Mediterranean coast runs along the bottom with a curious westward curve at its north end. It covers a relatively limited area, and extends only to Sidon and Damascus on the north, to the mountains beyond the Jordan on the east and to Gaza and the end of the Dead Sea on the south. The Jordan, other rivers and lakes are coloured dark olive between drawn outlines, and the same colour is used for the Mediterranean. Mountain areas are also drawn in outline plan; most are coloured pale olive, but a few are coloured beige. Towns are shown by drawings of a building, sometimes two; most have towers and battlements and many of the shapes verge on the fantastic. They are coloured red, and those that do not adjoin mountain areas have been placed on little hillocks, coloured pale olive. However, Jerusalem, which one might have expected to stretch fantasy to its limits, is marked only by an unnamed cross in a circle; this is coloured mauve, like details on symbols for individual monuments, such as the strange pepper-pot tower that marks the tomb of Job north of the sea of Galilee. All the other towns are named, however, and so are many of the rivers and mountains, and throughout the map are notes mostly of biblical events connected with the particular places. The twelve tribes are all named in large letters with a splash of red, and their boundaries are drawn in red. At the left end of the map is a list of nearly two hundred place-names, arranged alphabetically by first letter only, naming the tribe to which each belonged – or else placing them beyond the Jordan or beneath the Dead

52 It is worth noting how the short version of the text shows that the omission of a few words in the manuscripts of the full text that Laurent used has moved the pictured marble from the walls of the church of the Nativity at Bethlehem to the floor (*Burchardus, Liber de descriptione*, p. 114; *Peregrinatores*, ed. Laurent, p. 79).

53 I am grateful to Dr Stefania Ricci and Dr Simone Sartini for this information.

54 *Elenco dei documenti orientali e delle carte nautiche e geografiche chi si conservano negli archivi di stato di Firenze e di Pisa* (Florence, 1878), pp. 29-30.

55 Röhricht, 'Karten und Pläne I'.

56 It is interesting that the mapmaker has twice used arabic figures instead of roman: '24^or' in a note beside the upper Jordan (see Appendix 1 on p. 105) and '38' in a note at the south end of the map.

57 Desimoni, 'Carta della Terra Santa'.

58 Tesi, *Monumenti*, p. 25; there is a passing reference to the map in Degenhart and Schmitt, 'Sanudo und Veneto', p. 76.

Sea. Not all the names on the map are in the list,[59] and it is not clear why some were included, others left out.

The outlines of coast, rivers, mountain areas and the hillocks below the town symbols were all first drawn in pencil. These drafting lines are especially clear at the south end of the map, while in some places, such as the middle Jordan, the overdrawn ink is now so pale that the pencil is the effectively visible outline. Some at least of the town symbols were drafted in pencil, as at Pella and Tiberias, but others may have been drawn freehand in ink. Inking over the drafting lines was the next stage in the map's production. At *fons Engaddi*, west of the Dead Sea, the ink lines diverge from the pencil draft and at three points river courses have been altered after the ink lines were drawn but before colour was added: in the south-west corner of the map, north of Emmaus and, most startlingly, south of Mount Tabor, where the rivers Kishon and *Cadimin*, originally separate, were now linked, seemingly giving the sea of Galilee an outlet to the Mediterranean **[51]**. The dull and pale olive colours, and the red for the town symbols, were next added, and the tribal boundaries, already drafted in pencil, were overdrawn in red ink. The wording was written after water, mountains and towns had been coloured and the tribal boundaries inked, and the list of places at the left end of the map seems to have been entered last of all.

The map may have been left unfinished. South and west of the sea of Galilee, three mountain areas and three hillocks below towns have had lines of shading drawn in, and these mountain areas and one of the hillocks have been coloured beige **[52]**. It seems likely that this was to have been done throughout the map but that for some reason the operation was halted; some mountain areas and hillocks lack even the basic

59 E.g. Jerusalem and Toron are omitted.

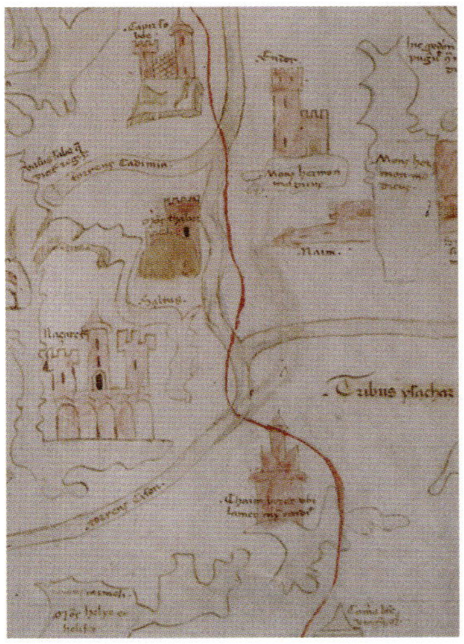

51. Running correction on the map. A branch of the river Kishon was redrawn while the map was being copied, producing a waterway that links the sea of Galilee with the Mediterranean. Florence, Archivio di Stato, Carte nautiche, geografiche e topografiche 4.

52. Shading on the map's relief features. Shading added to mountains and to the hillocks below towns was perhaps intended to be applied throughout the map, but in fact was restricted to this small area. Florence, Archivio di Stato, Carte nautiche, geografiche e topografiche 4.

pale-olive colouring, though it may simply have been lost through fading. Again, details are picked out in mauve on some monuments but not on all and this too may point to incomplete work. Apart from this, as Röhricht noted, later hands are responsible for a few of the inscriptions; they are the work of two writers, one of them perhaps as late as the fifteenth century, and are listed in Appendix 2 on p. 106.[60]

The map is a copy of an existing map, now lost, not an original production. On it are faint traces of a grid of vertical and horizontal lines, presumably to enable the outlines to be copied the more accurately; they were drawn in pencil, then erased more or less completely. Some vertical and horizontal lines are just visible on the two left-hand membranes, none at all on the third, and only a series of horizontal lines on the right-hand membrane – perhaps in this uncluttered part of the map these horizontal lines were all that were needed for accurate copying **[53]**.[61] The lines were carefully and precisely ruled, 14.5 mm apart on both axes. There is further evidence that the map was copied from another – and copied by someone with limited knowledge of the area – in the mistaken place-names that both Röhricht and Desimoni noted; Gath appears as *Thet*, Caymon as *Chaimborat*, and so on.[62] Again, the tribe of Gad has been placed to the south of the tribe of Reuben instead of to the north, while although the half-tribe of Manasseh this side of the Jordan is so described there is no sign of the half-tribe beyond the river; that the map's note beside Heshbon places it in the tribe of Reuben, though it is unequivocally in the area the map allocates to the tribe of Gad, suggests that the two names were reversed in making this or a preceding copy, and as the list at the left-hand end follows the map when it identifies places as in the tribe of Gad or Reuben, it must have been compiled then or at a later stage in transmission.[63] For all the care that was clearly taken over the outlines, these other

60 Röhricht, 'Karten und Pläne I', p. 9, saw only one later hand, but there are certainly two, and he identified some, but not all, of the additions on his reproduction.

61 In the Archivio di Stato at Florence I copied the positions of the lines on to a copy of Röhricht's redrawn version of the map; from this I later entered them on a photographic copy of the map, whence they have been copied on to a digital image to make fig. 53. While every care was taken at each stage, it is unlikely that total precision has been achieved, and fig. 53 should not be relied on for the exact positions of the lines.

62 Röhricht, 'Karten und Pläne I', p. 10; Desimoni, 'Carta della Terra Santa', pp. 248-9 (where some spellings of place-names from the map are faulty).

63 It may well have been made for this particular copy; the list, like the map, has *Betron* for *Bosra*, but this mistake does not occur on the Pierpont Morgan map (see chapter 11, pp. 128-32), which likewise reverses the tribes of Gad and Reuben.

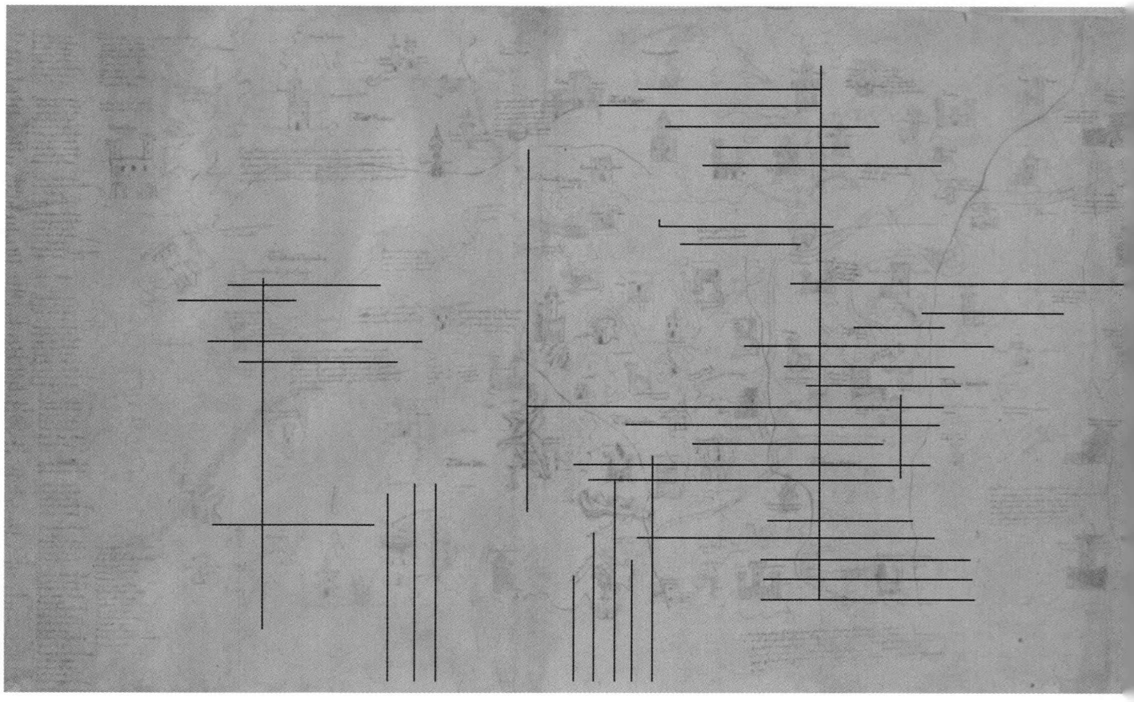

details of the map seem to have been less exactly reproduced – though the fault may of course have originated not in the present map but in an exemplar that was itself an erroneous copy. We might imagine – but this is the merest guess – that the original map from which our copy derives, whether at one or more removes, was a more polished production, with hills shaded throughout, with town symbols carefully drawn and with a representation of Jerusalem so elaborate that the copyist's skill or courage failed him and he replaced it with the simple cross in a circle. But even if we have no more than a shadow of what once was, it is a magnificent map and of great interest and importance.

Moreover its outlines of coast, rivers and mountains seem to be entirely the original mapmaker's own work. It is a map based on very considerable local knowledge. We have seen how our copyist altered rivers in the course of his work, notably in linking the sea of Galilee with the Mediterranean; we may suppose that, himself much less well informed, he took these from another map that he mistakenly thought was better – or, indeed, from Burchard's text, which gives the source of the Kishon two outlets, one to the Mediterranean, one to the sea of Galilee.[64] The original mapmaker will of course also have known where towns, villages and monuments lay, and how far he was guided by Burchard's text in choosing the places to show we cannot tell; where, as often, the map gives us no more than a symbol and a place-name, or the briefest reference to a biblical event, there is nothing to point to Burchard as the source. It is only when we look at the longer notes that the map's debt to Burchard becomes clear. They are not direct quotations from Burchard, but paraphrases that so closely echo Burchard's words that we are left in no doubt of their origin. Thus at the *fons Phiala*, north-east of the sea of Galilee, the map reads:

64 *Peregrinatores*, ed. Laurent, p. 48; Denys Pringle, 'The spring of the Cresson in crusading history', in *Dei gesta per Francos: études sur les croisades dédiées à Jean Richard: crusade studies in honour of Jean Richard*, ed. by Michel Balard et al. (Aldershot, 2001), pp. 237-8.

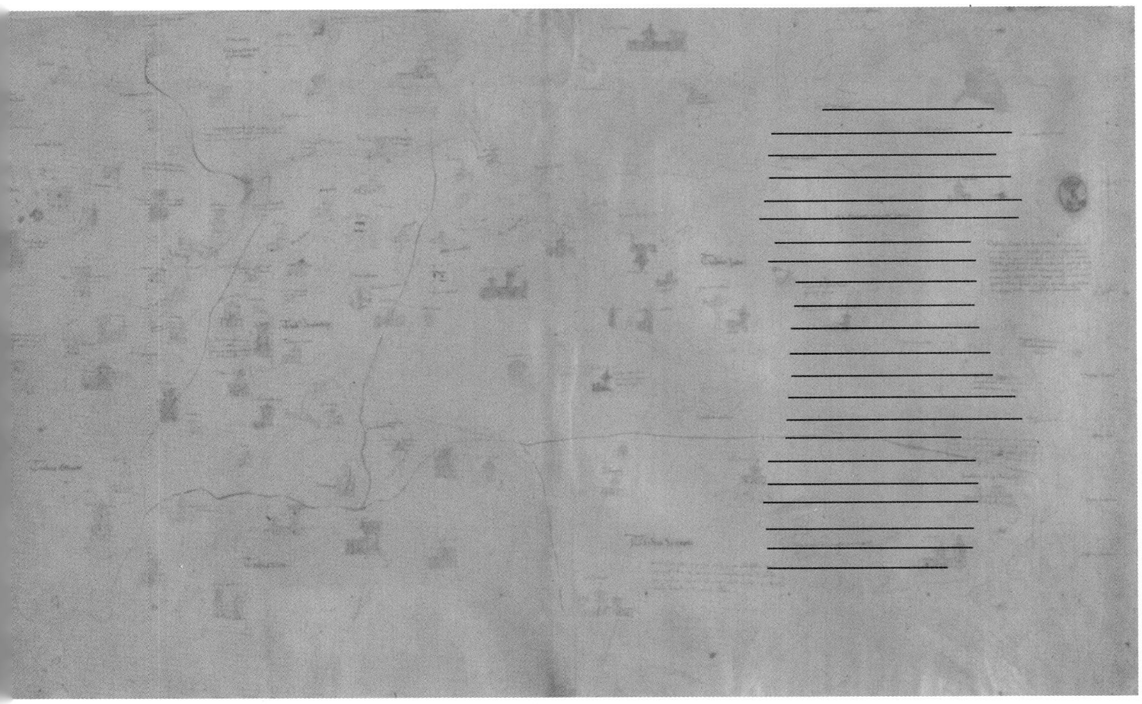

53. Traces of the erased copying grid. These ruled lines are barely visible on the map; the process of copying them may have introduced minor imprecision here. Florence, Archivio di Stato, Carte nautiche, geografiche e topografiche 4.

> Hunc fontem Josephus appellat Phialam, Marcus Dalmanicha, Mattheus Magedan, Saraceni Modin. Hinc est verus ortus Jordanis, unde palee hic misse recipiuntur in Dan subterraneo meatu ducte.

What Burchard tells us is:

> Dicit enim Iosephus , et uerum est, quod . . . est fons, qui appellatur Phiala, semper plenus et nunquam effluens, sed subterraneo meatu usque ad locum istum fluit et effluit ibi, et Dan uocatur. Quod per paleas in Phialam missas et in fonte Dan receptas sepius est probatum. Sarraceni autem fontem illum non Phialam, sed Medan uocant . . . Matheus uero locum illum Magedan appellat, Marcus Dalmanutha.[65]

Again, at Shiloh, north of Jerusalem, the map reads:

> In isto Silo mansit longo tempore tabernaculum federis et arca testamenti, nunc dicitur Sanctus Samuelis.

Here Burchard has:

> mons Silo, qui nunc ad sanctum Samuelem dicitur . . . In hoc loco fuit archa Domini longo tempore et tabernaculum federis, quod fecerat Moyses in deserto.[66]

Nearly all the place-names on the map are mentioned by Burchard and nearly all the longer inscriptions derive from his text. Nearly all – but not quite, though we must always remember that those missing may well be found in some version of the text that differs from Laurent's and that some of the map's place-names are unrecognisably corrupt.[67] The place-names include *Achalon* (north of Tyre) and the dwelling of the Virgin Mary (*camera beate virginis Marie*, north of Caesarea Palestina); among longer inscriptions is the note of David's crossing the Brook Besor (east of Beersheba). This last might well have been taken directly from the Bible,[68] and this was probably the case for a group of five notes referring to late stages of the Exodus: the altar of the children of Reuben, Gad and Manesseh and the land of King Sihon at Heshbon (either side of the middle Jordan), the sojourn of the children of Israel in the plain of Moab, with the death of Moses, and their passage across the Jordan (both north of the Dead Sea), and the long note on the desert of Kadesh (at the south end of the map).[69] As we might expect, the notes added by the two later hands, listed in Appendix 2, are unrelated to Burchard's text; he mentions the ravens that fed Elijah, but in quite different words.[70]

Can it have been a version of this map that Burchard sent to his friend at Magdeburg? Probably not. It is difficult to see how it could have been accommodated on a single sheet of parchment – though, as we shall see (chapter 10, p. 114), it was not long before a reduced version was produced – and for what it is worth the text refers to a single *pellis*, not *pelles* in the plural, nor even *rotulus*, a roll. Moreover, however much the map may have drawn on Burchard's text, it does not fully illustrate it; Sidon is the most northerly town shown on the coast, while Burchard goes on to describe Beirut, Byblos and other places as far north as Baniyas and the castle of Margat. But while it is unlikely that this was the map, it is not impossible, and we shall return to the question in looking at the small Burchard map (chapter 12, p. 146).

65 *Peregrinatores*, ed. Laurent, p. 32.
66 Ibid., p. 76.
67 Röhricht, 'Karten und Pläne I', pp. 10-11; Desimoni, 'Carta della Terra Santa', pp. 247-8, 251-2.
68 I Samuel 30.9-10.
69 Joshua 12.10; Deuteronomy 1-3, 34; Joshua 3.14-17; Numbers 13.
70 *Peregrinatores*, ed. Laurent, p. 57.

APPENDIX 1

QUESTIONED READINGS IN RÖHRICHT'S REPRODUCTION

The writing is faint, and some of the changes listed should be seen as a second opinion rather than as a firm emendation. The conventions used by Röhricht in transcribing have been followed here. The area of each tribe and identifying place-names are as given on the map itself.

List of names at left end of map
- Betrecha: *delete* tribu
- Basemut: *read* Besemut
- Emaus: *delete* civitas
- Hermon modicus mons: *delete* tribu
- Metada: *read* Mesada
- Sebaste: *delete* etiam

Beyond the left border of Röhricht's reproduction
- *Add* Syria damasci

Tribe of Reuben
- Note beginning 'Tota ista terra a monte Hermon': *for* Decapolis *read* Decapoleos
- Circle marking fons Phiala: *add* fons

Tribe of Naphtali
- At Aque Maron: *for* nimbis *read* niuibus
- Note beginning 'Hic Josue pugnavit': *for* reliquos *read* 24^{or}
- Note beginning 'Tota terra a monte Libano': *for* Tota terra *read* Ista terra, *for* Sophot *read* Sphot

Tribe of Asher
- *For* Vallis Senim *read* Vallis Sennim

Tribe of Gad
- At Manaym: *for* decollatus *read* decollatur

Tribe of Zebulun
- Between Neptatin and Abelina: *add* Dothaim *without town symbol*

Tribe of Issachar
- Note beginning 'Hic Gedeon': *for* Gedeon *read* Gedion
- At Suman: *for* mater *read* mulier

Tribe of Manasseh
- At Genni: *for* decem *read* x
- At Macedo: *for* Sullola *read* Subbola
- At Manacat: *for* nunc *read* que nunc etiam
- At Dora: *for* nunc *read* nunc asos nuncupatur

Tribe of Ephraim
- *For* Tirsa *read* Tersa
- *For* Pharael *read* Phasael
- At Samaria: *for* Samaria *read* Sammaria, *for* caput decem *read* capud x
- *For* Puteus Samarie *read* Puteus Sammarie

Tribe of Benjamin
- Note beginning 'In isto Silo': *for* ad Sanctum Samuelem *read* Sanctus Samuelis

Tribe of Dan
- At Thet: *for* iblis *read* iblim
- *For* Gath . . ha *read* Gathaha

Tribe of Judah

For Euqualiter *read* Engalim *(the name is written twice)*

For Beliga *read* Betiga

For Afa sonthamar *read* Asa sonthamar

Note beginning 'Hunc torrentem': *for* ceperant . . . *read* ceperant Sicelech

Tribe of Simeon

For Gaza *read* Gasa

Note beginning 'Tota terra ista': *for* ex tribu *read* de tribu

Beyond Jordan

Note beginning 'Terra siue regnum': *for* Osebon *read* Esobon

Note beginning 'Baptismus Christi': *for* capella *read* cappella, *for* baptistae *read* baptiste

In Dead Sea

For Gomorra *read* Gommora

For Achama *read* Achana

APPENDIX 2

INSCRIPTIONS ADDED IN LATER HANDS

The writing is faint and the ascription to different hands can only be tentative. The area of each tribe is as given on the map itself.

Second hand (14th century)

Tribe of Naphtali

Note beginning 'Cedos Neptalym'

Tribe of Judah

Note beginning 'Ista Bersabee'

Third hand (15th century)

Tribe of Reuben

'Syria Damaschi et terra Hamach, postea fuit terra sancta promissionis'

Tribe of Naphtali

'Introitus terre Hemach, que est Siria Damaschi'

'Introitus terre sancte promissionis'

Tribe of Ephraim

At Torrens Karuch: 'hic corvi pascebant Eliam' ('Torrens Karuch' is probably in the original hand)

Tribe of Judah

'In monte isto factus fuit primus homo'

Tribe of Simeon

Note beginning 'Tota terra ista'

Beyond Jordan

'Ermon'

THE GRID MAPS
1320–1339

10

A Vatican City, Biblioteca Apostolica Vaticana, ms. Pal. lat. 1362, ff. 7v–8r **[54]**. Parchment, 29 × 47 cm. Brown, blue and red ink, with brown (several shades), green and grey paint. Grid drawn in pencil. Executed at Venice, 1320.

B Oxford, Bodleian Library, MS Tanner 190, ff. 205v–206r **[55]**. Parchment, 28.5 × 44 cm. Brown and red ink, with brown (several shades), green and grey paint and gold leaf. Grid drawn in pencil. Executed at Venice with text dated [1321 × 1324].

C Vatican City, Biblioteca Apostolica Vaticana, ms Reg. lat. 548, ff. 141v–142r **[56]**. Parchment, 36 × 53 cm. Brown and red ink, with brown, yellow–brown and green paint. Grid drawn in brown ink. Executed at Venice with text dated [1322 × 1329].

D Florence, Biblioteca Riccardiana, ms 237, ff. 142v–143r **[57]**. Parchment, 36 × 51 cm. Brown and red ink, with beige, yellow–brown and green paint. Grid drawn in brown ink. Executed at Venice with text dated [1329].

E London, British Library, Add. MS 27376*, ff. 188v–189r **[58]**. Parchment, 34.5 × 51 cm. Brown and red ink, with brown, yellow–brown and green paint. Grid drawn in brown ink. Executed at Venice with text dated [1330].

F Brussels, Bibliothèque royale de Belgique, ms 9347–9348, ff. 162v–163r **[59]**. Parchment, 35.5 × 51 cm. Brown, blue and red ink, with yellow–brown and blue–green paint. Grid drawn in pencil. Executed at Venice with text dated [1331 × 1332].

G Brussels, Bibliothèque royale de Belgique, ms 9404–9405, ff. 173v–174r **[60]**. Parchment, 35.5 × 51 cm. Brown, blue and red ink, with yellow–brown and green paint. Grid drawn in brown ink. Executed at Venice with text dated [1331 × 1332].

H Paris, Bibliothèque nationale de France, ms. latin 4939, ff. 10v–11r (central portion of each page) **[61]**. Parchment, approximately 27 × 70 cm. Brown, gold and red ink, with brown (several shades), green, grey and yellow paint. Grid drawn in brown and red ink. Executed at Naples, [1329 or soon after].

I Vatican City, Biblioteca Apostolica Vaticana, ms Vat. lat. 1960, ff. 265v–266r (lower half of each page) **[62]**. Parchment, approximately 20 × 53 cm. Grey–black and red ink, with red–brown, olive–green and yellow paint. Grid drawn in grey–black ink. Executed at Naples, [1334 × 1339].

J Paris, Bibliothèque nationale de France, ms. latin 15363, ff. Iv–IIr **[63]**. Parchment, 36 × 52 cm (cut down). Dark grey and red ink. Grid drawn in dark grey ink. Executed at Naples (?), [14th century].

AFTER THE LOSS OF ACRE in 1291 there were many writers who urged a new crusade to restore Christian rule to the Holy Land.[1] Among them the most persistent, and probably the most influential, was Marino Sanudo. He was born

1 Two who, like Marino Sanudo, used maps in making their case, were Fidenzio da Padova, whose map of the eastern Mediterranean survives in two copies of his text, and Galvano da Levanto, whose map of the Holy Land has failed to survive (Gautier Dalché, 'Cartes, réflexion stratégique et projets', pp. 80–4, plates 1, 2).

about 1260 into a well-to-do merchant family at Venice and he knew the eastern Mediterranean well – he had indeed visited Acre before its fall. He wrote a book begun in 1306 (our earliest copy dates from about 1309) proposing an economic blockade of Egypt as a way of breaking the Mamluks' power over Palestine, but it was some ten or fifteen years before he expanded this into his *Book of secrets of those faithful to the Cross* (*Liber secretorum fidelium crucis*). With this work he initiated a serious campaign for a crusade. Between 1321 and 1332 he wrote to and visited many of the

most important people in western Europe, giving them copies of his book and encouraging them to take action. Preparations for an expedition were twice actually begun, but each time proved abortive, and by the time Sanudo died in 1343 nothing had happened.[2]

One of the first to receive Sanudo's book was Pope John XXII. Sanudo visited him at Avignon in 1321 and gave him two copies, along with a little atlas of ten maps: a world map, five portolan charts of the Mediterranean and the Black Sea, a map

2 Accounts of Sanudo's life and work are in Friedrich Kunstmann, 'Studien über Marino Sanudo den Aelteren mit einem Anhange seiner ungedruckten Briefe', *Abhandlungen der historische Classe der Königlichen Bayerischen Akademie der Wissenschaften*, 7 (1855), pp. 695–819; Degenhart and Schmitt, 'Sanudo und Veneto', pp. 3–16; Cardini, 'Per un' edizione critica', pp. 193–212; C.J. Tyerman, 'Marino Sanudo Torsello and the lost crusade: lobbying in the fourteenth century', *Transactions of the Royal Historical Society*, 5th series, 32 (1982), pp. 57–73; Leopold, 'Crusading proposals', pp. 216–21.

54. Grid map A, drawn at Venice, 1320; east at the top. The only grid map that does not accompany text, being in an atlas of ten maps that Sanudo gave to Pope John XXII. Vatican City, Biblioteca Apostolica Vaticana, ms Pal. lat. 1362, ff. 7v–8r.

of the eastern Mediterranean with lands to the east, a map of Palestine and plans of Jerusalem and Acre. The atlas survives in the Vatican library (the map of Palestine is here map A) and so does one of the two copies of the *Book of secrets*, which also

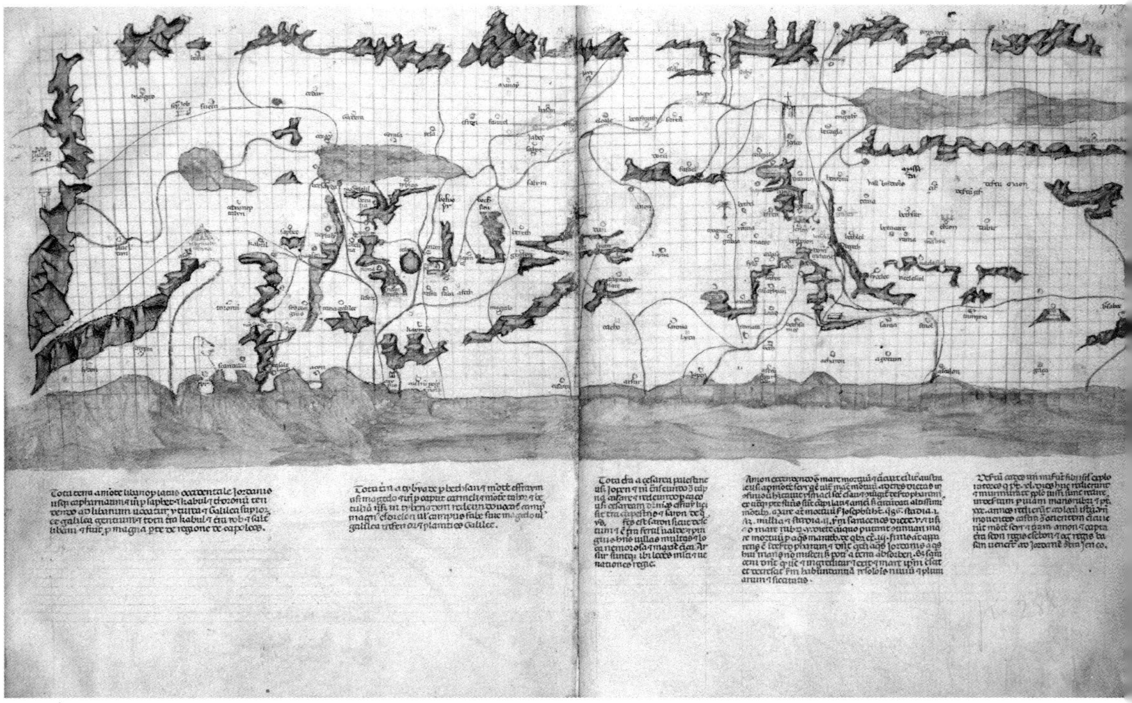

55. Grid map B, drawn at Venice, 1321 × 1324; east at the top. Unlike the other grid maps there are no notes from Burchard's text, the tribes are not named (though their boundaries are shown) and the map is integrated with Sanudo's text, not on a separate quire of maps. Oxford, Bodleian Library, MS Tanner 190, ff. 205v-206r.

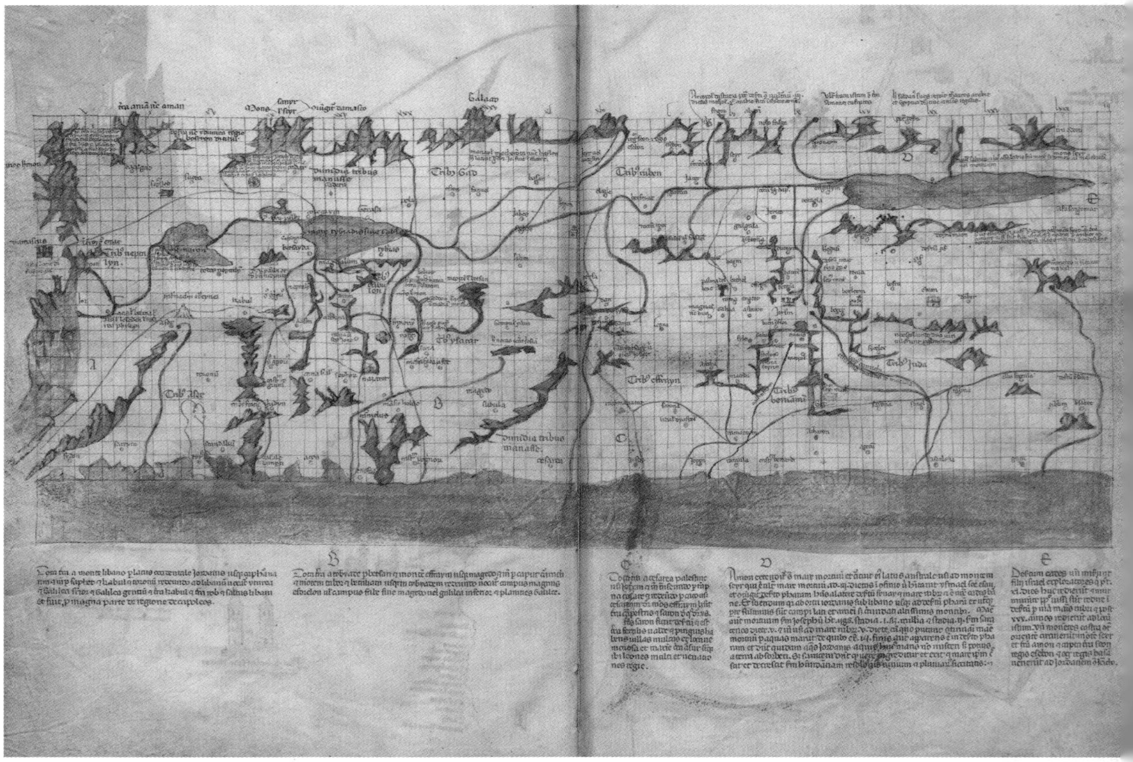

56. Grid map C, drawn at Venice, 1322 × 1329; east at the top. This was the earliest of the grid maps to be reproduced in print, from an engraved plate in 1611. Vatican City, Biblioteca Apostolica Vaticana, ms Reg. lat. 548, ff. 141v-142r.

contains the world map, four charts and the map of the eastern Mediterranean.[3] Seven other copies of the book survive; one, probably earlier than John XXII's copies, has only the map of the eastern Mediterranean, but all the others include the map of

3 Degenhart and Schmitt, 'Sanudo und Veneto', pp. 68–9, 105; Leopold, 'Crusading proposals', p. 221.

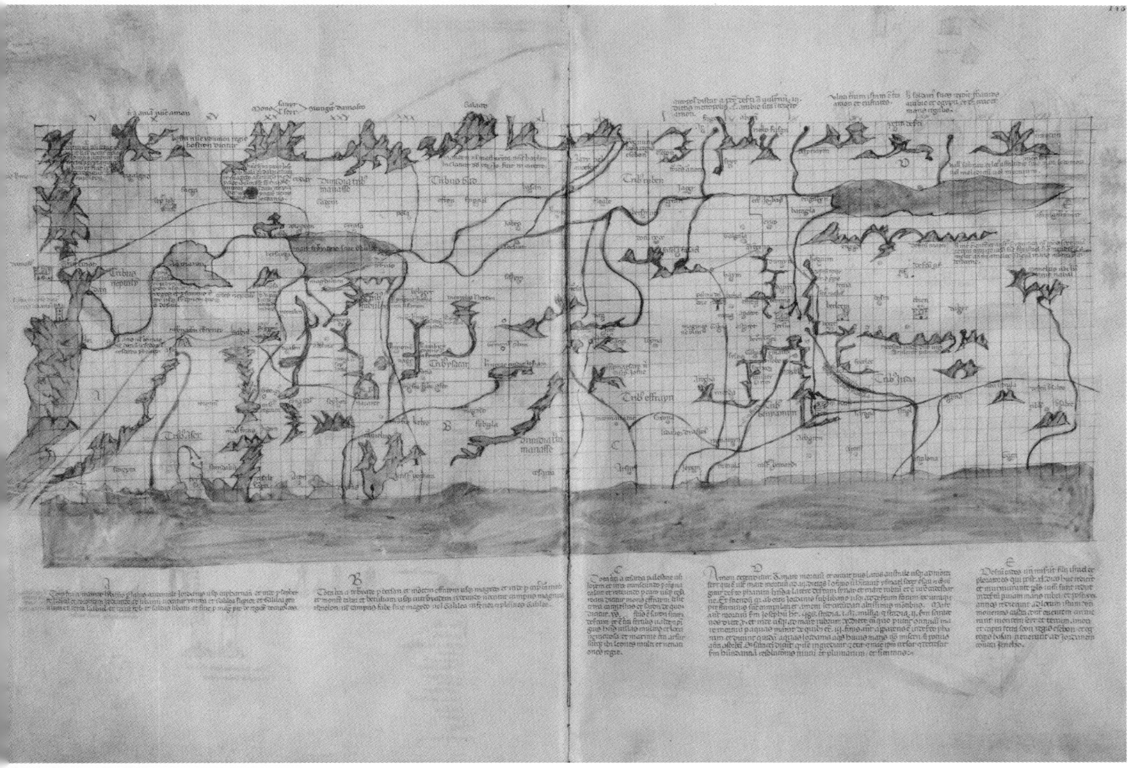

57. Grid map D, drawn at Venice, 1329; east at the top. The wording on grid maps C, D and E is in the same hand. Florence, Biblioteca Riccardiana, ms 237, ff. 142v–143r.

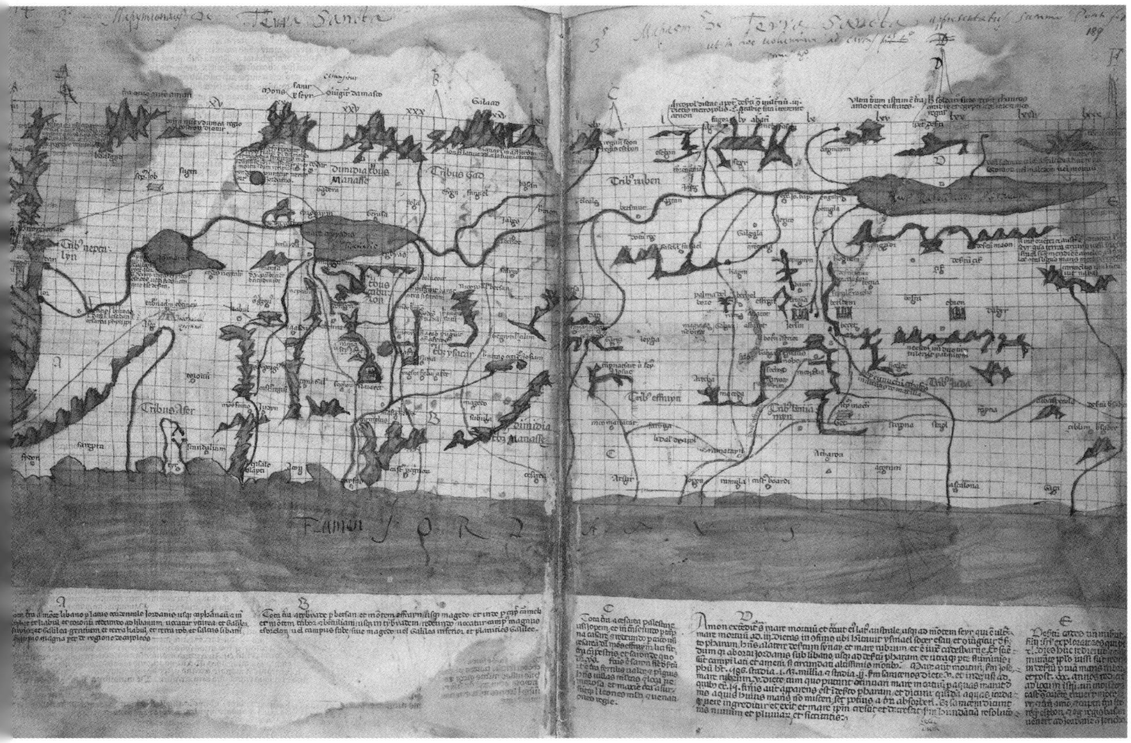

58. Grid map E, drawn at Venice, 1330; east at the top. The radiating lines on the sea in the lower right corner were accidentally transferred from another map in the course of conservation. BL, Add. MS 27376*, ff. 188v–189r.

Palestine (here maps B to G) along with copies of three, four or, in one case, seven of the other maps that are in the atlas. Some recipients may have been given maps alone, without the book; Sanudo wrote how, after presenting the book and maps to the king of France, 'I at once hurried to the house of the lord Louis, count of Clermont, giving him maps and some other items concerning these affairs.'[4] As Patrick Gautier Dalché has suggested, the maps played an important part in Sanudo's argument, enabling his strategy to be easily visualised at successive levels of detail.[5]

Some of the manuscripts are dedicated to one or other of the important persons to whom Sanudo gave copies of the book. The manuscripts containing maps F and G begin with a dedicatory letter of 1332 to Philip VI, king of France, and these were the copies actually given to him, for we know their successive owners from the early fifteenth century onwards. The manuscript containing map B has a dedication to Robert, count of Boulogne and Auvergne from 1313 to 1324, but in this case we cannot be sure that it was not simply copied from the book that was presented. More help in dating is provided by the copies of letters to and from Sanudo that are appended to the text in the manuscripts containing maps C to G: the later the manuscript, the later the letters that are included. However, in every case the letters are on a separate quire that need not have been written at the same time as the text to which it is attached; their dates are only contributory to the date of the text. This is taken into account by Bernhard Degenhart and Annegrit Schmitt, and it is their covering dates for each copy of the text that are given in the list above.[6] However, a word of caution is in order. The maps in the manuscripts that include maps C to G are in each case, like the letters, on a quire by themselves; this quire follows the quire with the letters in the manuscript containing map E but precedes it in the others. This means that the maps too may not be of the same date as the text. A supply of quires with the maps drawn on them might, for instance, have been prepared in advance to be added to copies of the text as they were written; we should remember that many more copies of the book will have been written than have survived, and we may note that maps C, D and E are the work of one hand, maps F and G the work of another. This does not apply to map B; in this manuscript the maps occupy later leaves of the same quire as the end of the text and we may reasonably suppose that they were drawn at the same time.[7]

Nor, of course, does it apply to map A in the atlas of ten maps, drawn on a single quire of ten leaves. It was Konrad Kretschmer who in 1891 first showed how important this atlas is to our knowledge of Sanudo's maps. On one of its portolan charts is the note 'Pietro Vesconte of Genoa made this map or [. . .] in the year of the Lord 1320'.[8] The missing word starts *qua* (the rest is cut off) and it may well have been a form of *quaternam*, quire. In fact six of the atlas's maps are all or partly in Vesconte's hand; the map of Palestine is not one of them, but its left-hand page is on the same bifolium as the map Vesconte signed and we need not doubt that all ten maps in the atlas were drawn in his workshop and at the same time. Nor need we doubt that maps B to G, and the other maps that accompany each of them, all came from the same source, though Vesconte's hand reappears only in five of the maps that accompany map E.[9]

Of Pietro Vesconte we know only what we learn from his maps. Four others

4 'postquam praesentaueram librum *Fidelium Crucis* & Mappas mundi domino *Regi Franciae*, immediatè acceleraui ad domum domini *Ludouici comitis Claremontis*, eidem praesentans Mappas mundi, & quaedam alia de negotiis supradictis', letter to the bishop of Mende, 1326 (*Gesta dei per Francos*, ed. Bongars, II, p.296).

5 Gautier Dalché, 'Cartes, réflexion stratégique et projets', pp. 86-7.

6 Degenhart and Schmitt, 'Sanudo und Veneto', pp. 12-13, 16. Invaluably, Degenhart and Schmitt provide on pp. 21-7 details of each manuscript containing the relevant work of Sanudo and of Paolino Veneto, and on p. 105 a table showing what maps are in each.

7 In MS Tanner 190, containing map B, ff. 200-208 are a single gathering of ten leaves (there is a blank unnumbered leaf between ff. 202, 203); the end of Sanudo's text occupies ff. 200r-202v, a text 'De insulis minoribus' has been added on f. 203r and the maps, followed by a table of rulers in the Near East, are on ff. 203v-208r.

8 'Petrus Vessconte d' Ian[. . .] fecit istam Cartam uel qua[. . .] anno domini M°ccc°.xx.' (Vatican City, Biblioteca Apostolica Vaticana, ms Pal. lat. 1362, f. 3v). Further digits may have been cut away from the end of the date, but on other grounds it seems likely that it really is 1320.

9 Kretschmer, 'Marino Sanudo'; Degenhart and Schmitt, 'Sanudo und Veneto', pp. 64-7.

survive that he signed with dates from 1311 to 1318, all of them sea-charts, and some others have been identified as his work, like those accompanying map E; here, significantly, it is the charts that are entirely in his hand, while he contributed just the Mediterranean to the world map and the maritime portion to the map of the eastern Mediterranean with lands to the east. Another atlas and chart are signed by Perrino Vesconte with the dates 1321 and 1327; this may have been Pietro under another name – Perrino is a diminutive of Pietro – but was more likely a younger relative, perhaps a successor to the business. While Pietro described himself as of Genoa, it is clear from several of the inscriptions that his workshop was in Venice. He is an important figure in the early development of the portolan chart, and though his surviving maps are few they point to a considerable output. We get the impression of a cartographer who himself drew charts, in which he took a strong personal interest, but who also ran a workshop or studio that undertook other kinds of graphic work.[10]

Vesconte was of good family; he and Sanudo may well have moved in the same circles in Venice and it is not surprising that Sanudo's work was illustrated by Vesconte's maps.[11] Nor is it surprising that some of these maps reappear in four manuscripts of the *Great chronology* (*Chronologia magna*) by Paolino Veneto, two of them including the map of Palestine (maps H and I). Paolino was a contemporary of Sanudo.[12] He became a Franciscan friar in 1293 and is sometimes called Paolino Minorita, the Friar Minor or Franciscan. From 1301 he worked for his Order in Venice, and he may have made a journey to the Holy Land at some point between 1316 and 1320.[13] Between 1314 and 1325 he undertook successive missions on behalf of the Venetian government, coming into contact with Robert of Anjou, king of Naples, both at Naples itself and, in 1321, at Avignon. There he was a member of the papal commission that considered the proposals of Sanudo, who was at Avignon at the same time. In 1324 Paolino was elected bishop of Pozzuoli, a few miles from Naples, and he served as councillor to the king until Robert's death in 1343. Paolino himself died the following year.

His *Great chronology* was a universal history set out partly in diagrammatic form, with tables showing pictures of rulers and so on. No two of the four manuscripts are identical: the work is abridged or enlarged, and the latest events mentioned in each give clues to the date of the copy. Maps fitted naturally into this context, being simply a particular form of diagram and in three of the manuscripts they are dispersed throughout the work; only in the manuscript containing map I are they brought together in a single quire at the end of the volume. This manuscript includes, along with the *Great chronology*, a different version of Paolino's history, set out in prose and entitled the *Satiric history* (*Satyrica historia*), and it is interesting that later manuscripts of the *Satiric history*, without the *Great chronology*, have no maps.

Paolino knew Sanudo well and it is likely that he knew Vesconte too.[14] Our earliest copy of the *Great chronology*, dating from 1327 or soon after, is in Paolino's own hand and the style of the illustrations shows that it was produced in Venice; its map of the lower Po valley and its plans of Antioch, Acre (newly revised), Jerusalem, Rome and Venice itself are probably the work of Vesconte's workshop. Another manuscript, produced soon after at Avignon, contains an abridged version of the *Great chronology* translated into Provençal and only two maps, the town plans of

10 Degenhart and Schmitt, 'Sanudo und Veneto', pp. 64-7; Tony Campbell in *History of cartography*, pp. 406-7; Pujades i Bataller, *Les cartes portolanes*, pp. 489-90.

11 Tony Campbell in *History of cartography*, p. 434.

12 There is a short account of Paolino's life and work in Degenhart and Schmitt, 'Sanudo und Veneto', pp. 16-20, 25-7; Ghinato, *Paolino da Venezia*, is a fuller account.

13 Ghinato, *Paolino da Venezia*, pp. 32-3.

14 Ibid., pp. 54-6.

Antioch and Jerusalem; the place where it was made seems to have determined the choice of maps in each copy of Paolino's work. The two that include the map of Palestine were made at Naples; they also include the maps of the world and of the eastern Mediterranean with lands to the east that were in manuscripts of Sanudo's work, as well as the town plans of Antioch, Acre, Jerusalem and Rome. The manuscript containing map I, dedicated to King Robert of Naples, also includes the map of the lower Po and maps of the whole Italian peninsula.[15]

As we shall see, there are few differences between these nine maps of Palestine, maps A to I; they are all versions of a single map, though it seems likely that two slightly different exemplars were used, one at Venice, the other at Naples. Of the map's source there is no doubt: it is a reduced version of the large Burchard map. It covers exactly the same area: with the coast at the bottom it extends to Damascus on the north, to the mountains of Gilead and Petra on the east and to the end of the Dead Sea on the south. The basically straight coastline has a similar pattern of indentations, the river system is the same – the Jordan follows the same winding course – and so for the most part are the mountain ranges, so that even some of their detailed shapes have survived the processes of copying. The tribes are named (except on map B) and their boundaries are shown. There are as many place-names as on our earliest large Burchard map in the Archivio di Stato at Florence and, while the scriptural and other notes paraphrased from Burchard's text have been severely pruned, a significant number appear on all but map B. While, as we shall see, the nine maps show towns in different ways, none reproduces the hillock below each town symbol that is so prominent a feature on the Archivio di Stato map.

The colouring is broadly the same on all nine maps and must reproduce the colours of the exemplars. Water is green, mountains are brown with varied shading, the names of the tribes are red and so too (except on map H) are the tribal boundaries. The shades of colour entered here in the list of maps can only be a rough guide; only in the Vatican library and the Bibliothèque royale in Brussels can any of the maps be compared side by side, and colour photographs can be misleading.[16] However, it is worth noting that on the two maps at Brussels, both written by the same hand, the blue-green used for water on map F is quite different from the yellower green on map G.

While the map is a reduced version of the large Burchard map, it was certainly not compiled from the Archivio di Stato map, the earliest copy that survives. We have seen that this is indeed an imperfect copy: names of tribes are misplaced and place-names miscopied. These mistakes do not recur on the grid maps, which must derive from a better copy. On the grid maps the tribes are correctly placed – Reuben is to the south of Gad, not the north, and the half-tribe of Manasseh east of the Jordan is named – and, for instance, *Gerasa* and *Arsur* replace the Archivio di Stato map's *Gorsa* and *Dora*. On the other hand a very few names are spelled better on the Archivio di Stato map than on the grid maps; thus Timnath-heres, the burial-place of Joshua, Thamnathsare in the Vulgate and *Thanathesare* on the Archivio di Stato map, has on the grid maps become *Tampnatsore*, *Tampnathsare*, *Tampnatsare*, *Tamnatasere*, *Tampnatasere*.[17] Though the Archivio di Stato map and the grid maps both name about 150 towns and other sites, those named are not identical and some

15 Degenhart and Schmitt, 'Sanudo und Veneto', pp. 25-6, 83, 105 (where, however, the map of Acre in Paris, Bibliothèque nationale de France, ms latin 4939, f. 113v, has been overlooked). On the maps' contribution to our knowledge of Acre's medieval topography see chapter 8, note 18.

16 The problems of defining and comparing colours are discussed in P.D.A. Harvey, 'Colour in medieval maps', in *Signs and symbols: proceedings of the 2006 Harlaxton Symposium*, ed. by John Cherry and Ann Payne, Harlaxton Medieval Studies 18 (Donington, 2009), pp. 42-52.

17 Judges, 2.9; *Peregrinatores*, ed. Laurent, p. 28. The successive forms are on maps A; B; C-G; H; I.

twenty-odd names occur on one but not the other. Thus we see *Gloria in excelsis* (where the angels appeared to the shepherds), *Achalon* and Zeboyim only on the Archivio di Stato map, Scandelion, Ramla and the house of Zacharias only on the grid maps. It may be that the original large Burchard map included all these names and that some or others were left out when our copy and the exemplars of the grid maps were compiled. Alternatively, names may have been added – or omitted – at any

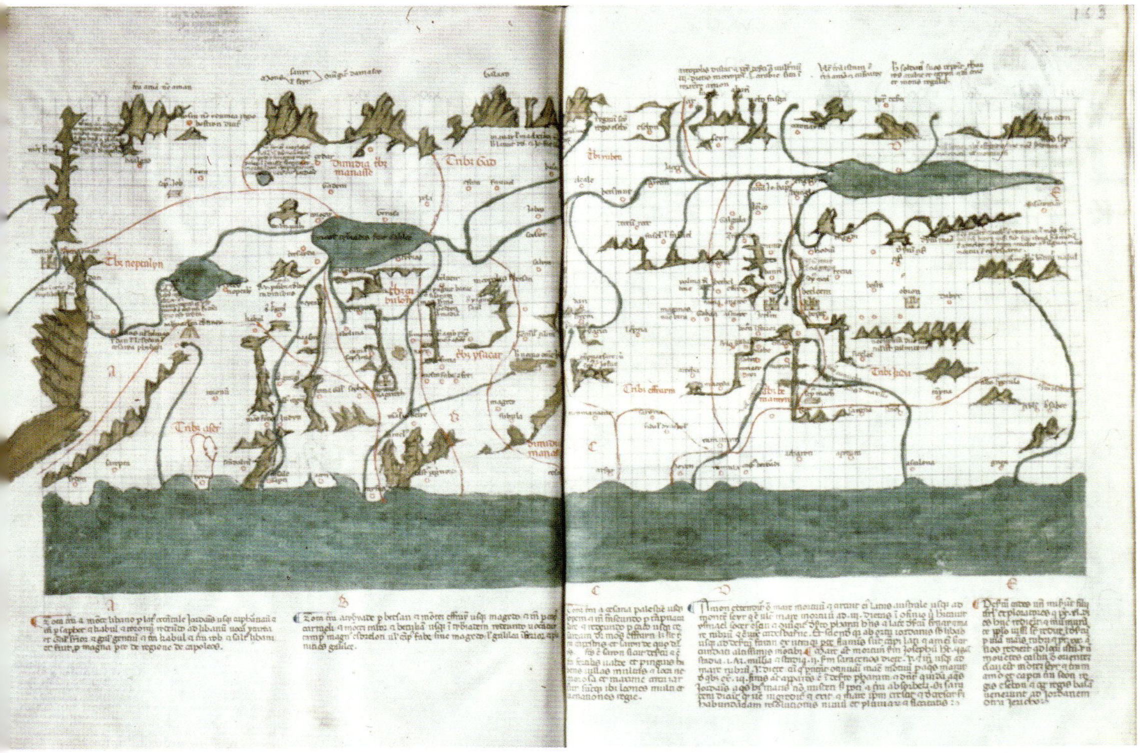

59. Grid map F, drawn at Venice, 1331 × 1332; east at the top. While the wording on grid maps F and G is in the same hand, water on map G is painted in a yellower green than on map F. Brussels, Bibliothèque royale de Belgique, ms 9347-9348, ff. 162v-163r.

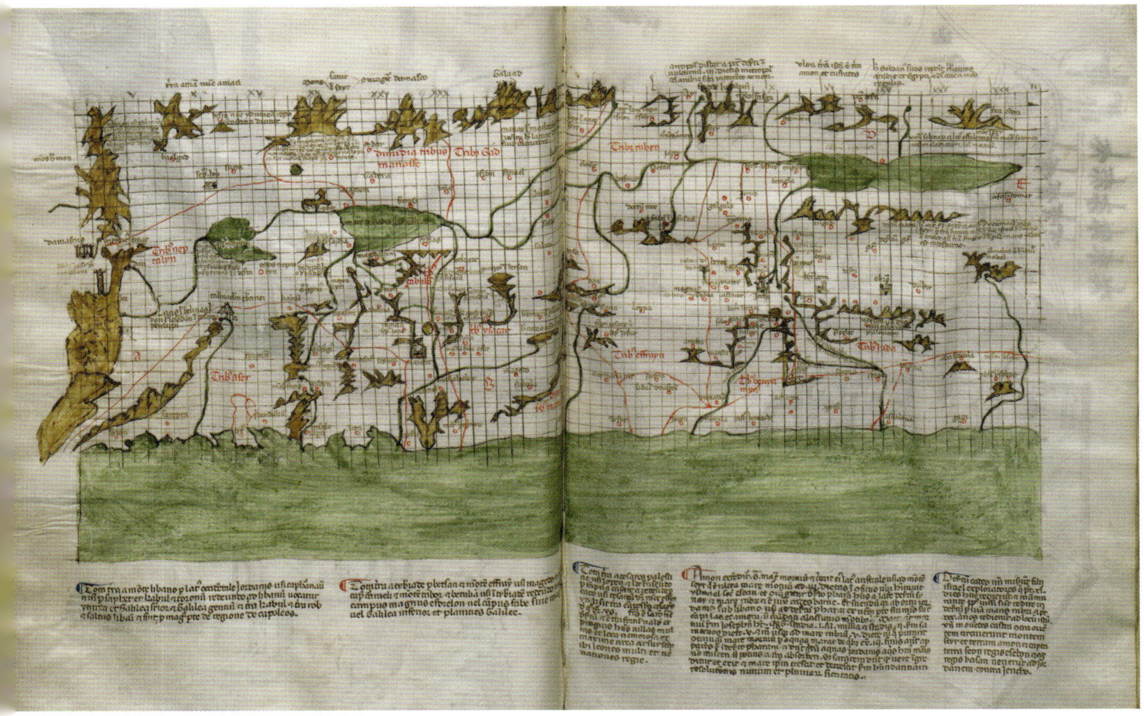

60. Grid map G, drawn at Venice, 1331 × 1332; east at the top. The manuscripts containing maps F and G were given by Sanudo to Philip VI, king of France. Brussels, Bibliothèque royale de Belgique, ms 9404-9405, ff. 173v-174r.

point in the process of transmission. Full analysis of the place-names on all surviving copies of the large Burchard map and on the grid maps, along with the texts of Burchard and Sanudo, might well tell us more about the derivation and composition of both.

The most striking feature of the grid maps is their overall grid of twenty-eight squares from east to west, eighty-three from north to south. It has intrigued many historians of cartography, for there is no parallel in contemporary or earlier European maps; it has been suggested that Vesconte may have seen grid-based Islamic maps and that these in turn may owe something to contemporary Chinese grid maps.[18] This is not impossible, but the grid, largely erased, that was used in copying the Archivio di Stato map points to an explanation nearer home. Vesconte or Sanudo realised that just such a grid might be used not just to copy the map accurately but to provide a guide to the places on the map. Sanudo sets this out in the later versions of the *Book of secrets*: 'Let us therefore imagine the Promised Land divided by lines into twenty-eight strips (*spatia*) extending from Mount Lebanon to the desert that leads into Egypt and into eighty-three strips drawn with lines crossing the others from west to east, so that we have a great many squares, each of one league or two miles.'[19] He then plots the position of each place on the map in terms of these squares, proceeding systematically strip by strip from east to west: 'Thus in the first strip at the top, in the east, moving from north to south, Ar, now Arcopolis or Petra, is in square 53, Petra of the desert or Montréal in 76. . . . In the second strip Bosra, now Idumea, is in square 12. In the third strip Heshbon is in square 50, Othonaym in 61. In the fourth strip Macheronta or Manaym, now Haylon, where David hid and John the Baptist was put in prison, is in square 36, Ayr or Auoth Seyr in 44, Sethyn in 55' and so on.[20] The idea was innovative and imaginative, and we can see a first step towards it in the list of places, naming the tribe where each one lay, on the Archivio di Stato map.

From the grid lines visible on the Archivio di Stato map it looks as if there too were 28 strips running north and south, but its proportions differ from the grid maps and there must have been about 108 strips running east and west. There is no obvious source for Sanudo's statement that the Promised Land was 28 leagues or 56 miles – that is, 28 strips – from east to west nor, indeed, for any overall measurement on this axis. On the other hand, St Jerome wrote to Dardanus in 414 that the distance from Dan in the north to Beersheba in the south was 160 miles, and this was copied by later writers from Bede onwards.[21] This does not precisely match the positions of the two towns on the grid maps, for Dan is in column 6, Beersheba in column 80, but it could all the same have been Sanudo's source for his 83 leagues or 166 miles from north to south – in some manuscripts Jerome's 160 was miscopied as 165.[22] We might guess – no more – that Vesconte or Sanudo took the 28 strips from the grid already used in making copies of the large Burchard map, but altered the lines on the other axis to reflect Jerome's figure, thus bringing the map into the right proportions for the double-page opening it was to occupy in successive volumes.

Whatever the origin of the grid, we are left with a question that is not difficult to answer. The list of places that was apparently the grid's *raison d'être* is set out in the later versions of Sanudo's *Book of secrets*; there is no such list in Paolino's *Great chronology*, the text with maps H and I, yet on these two maps the grid still appears.

18 Joseph Needham, *Science and civilisation in China* (7 vols, Cambridge, 1954-2004), III, p. 564, plates LXXXVII, LXVIII; P.D.A. Harvey, *The history of topographical maps: symbols, pictures and surveys* (London, 1980), pp. 144-6. Cf. Gautier Dalché, 'Cartes de Terre Sainte', p. 602n.

19 'Imaginemur itaque Terram Promissionis diuidi per lineas in XXVIII. spatia, quae protenduntur à monte Libano vsque ad desertum quo itur in Ægyptum; et in LXXXIII. spatia: tractis lineis super priores transuersaliter ab Occidente in Oriens, ita quòd resultent, ad quantitatem vnius leucae siue duorum milliarium, plurima loca quadrata' (*Gesta dei per Francos*, ed. Bongars, II, p. 246).

20 *Gesta dei per Francos*, ed. Bongars, II, pp. 246-9; Degenhart and Schmitt, 'Sanudo und Veneto', pp. 22-3.

21 *Patrologiae cursus completus, series latina*, ed. by J.-P. Migne, 22 (Paris, 1845), p. lxxxvii, col. 1104; *Sancti Eusebii Hieronymi epistulae*, ed. by Isidor Hilberg (3 vols, Vienna and Leipzig, 1910-18), III, pp. 169-70.

22 As we see in some manuscripts in Bede, 'De locis sanctis', p. 261; or in the text of Peter the Deacon, in *Itinera*, ed. De Sandoli, II, p. 180.

We may reasonably suppose that the grid was by this time seen as so much a part of the map that it would be copied mechanically without thinking of its function, except as an aid to copying. Earlier it had appeared on map A, in the atlas without text, perhaps because the little collection of maps, on a single quire, had been prepared with a view to adding it to a copy of Sanudo's text, perhaps again through mechanical copying of the exemplar. On map A, but also on maps B and F, the grid is drawn in pencil, not ink, but unlike the pencilled grid on the Archivio di Stato map it is heavily drawn, clearly with no intention of removing it by erasure.

There is much variation in detail between the nine maps A to I, mostly accidental from faulty copying. However, there are some consistent differences between the maps drawn in Venice (A to G) and those drawn in Naples (H and I), pointing to the use of two different exemplars, the master-copies from which the individual maps were copied. Thus at Tyre all the Venice maps show, unnamed, the four springs of the well of living waters, in the same form as on the large Burchard map with two streams flowing to the sea,[23] but they appear on neither of the two Naples maps. On both the Naples maps is a note to the west of the Dead Sea that is on none of those from Venice, about the balsam plant that Cleopatra brought to Egypt; this comes from Burchard's text and while it does not appear on the Archivio di Stato map this simply reminds us that this copy is not a perfect one.[24] Again, both the Naples maps give Megiddo south of Nazareth two names, *Magedo* and *Sububa*, corresponding both to Sanudo's list, where it is 'Magedo uel Sububa' and, with variant spellings, to the *Macedo nunc Subbola* of the Archivio di Stato map;[25] all the Venice maps, however, show two separate towns, one named *Magedo*, the other *Subula*. Other differences include the spelling of place-names; all the Venice maps call Acre *Acon*, for instance, the Naples maps *Achon*.

Some differences may simply point to different presentation, not to different exemplars. Unlike the Naples maps, all the Venice maps mark towns with a tiny red circle, nearly always in the centre of its square, though Jerusalem, Damascus, Nazareth and Hebron are marked with a building of one, two or (mostly) three towers, while at Bethlehem there is a church on maps C, D and G, a triple-towered building on maps E and F and a simple circle on maps A and B. Below each map are five long notes; they are lettered from A to E and letters on the map show the area each note refers to. Notes A, C and E are slightly altered versions of notes that appear on the Archivio di Stato map, describing the bounds of Upper Galilee, the area of Mount Ephraim and the desert of Kadesh. Notes B and D, one giving the bounds of Lower Galilee, the other an account of the land of Ammon, the Jordan valley and the Dead Sea, are not on the Archivio di Stato map, but we may reasonably suppose that they were taken from a more perfect copy. Map B differs in several ways from the other Venice maps. It confines its wording to simple place-names, omitting other notes, and it thus has a less crowded appearance. It omits the identifying letters from the notes below the map and from the map itself and while it shows the tribal boundaries it does not name the tribes. The buildings marking Jerusalem, Nazareth and Hebron are in gold,[26] while symbols for Damascus and some other features seem to be additions to the map, entered with a fine pen; among these are Haifa, the house of Zacharias and some other sites marked not with a red circle but with a two-pointed

23 Cf. Burchard's text, *Peregrinatores*, ed. Laurent, p. 24. Mysteriously, on maps C, D, F and G they, alone among rivers, are coloured red; there may have been some ambiguity in the colouring of the exemplar.

24 *Peregrinatores*, ed. Laurent, p. 61.

25 *Gesta dei per Francos*, ed. Bongars, II, p. 249.

26 The brown on the symbols for these places on map A may possibly be the ground for gold leaf that has flaked off.

pueri sui lauisse dicitur. Ad huc fontem in epiphania saraceni et christiani conueniunt et ibi ex deuocione corpora sua lauant. Ibi est aliud stupendum quia boues aquam predictam haurientes ad ortum [illegible] a uespera diei sabbati usque ad secundam horam diei dominice et si extorquati debent aquam non trahent. ·A· Babilone ad v leucas sunt quedam pyramides triangule mire alte que dicuntur fuisse horrea ioseph et inter illas ruine ciuitatis tebe ad ij leucas. Cui adiacet desertum tebaidos ubi fuit copia antiquitus monachorum. Supra babiloniam uero nilus descendit totus collectus et ibi latitudo egypti que nulla est [illegible] montuosa. Ascendit autem a loco dicto syene per millia ccc l et uenit usque meroen prius dicta saba ubi inclusit moyses ethyopes sicut milia

cere. Et est syene sub tropico estiuali ita ut suo tempore umbram proiciat contra austrum. Ortus autem nili sciri non potest nisi usque ad montes qui sunt a sinistris nubie per quos descendit et est ulterius locus inmeabilis. Terra que egypti alluitur que a nilo magis pene inaccessibilis est. Et parte noui occidentis confinia habet cum quadam prouincia barbarie que Barcha uocatur et est desertum medium xv dietarum. Ab austro est desertum vij dietarum et ultra usque nubiam. Ab oriente desertum thebaydos usque ad mare rubrum per iij dietas ad locum qui Beronice dicitur et est portus egypti in litore maris rubri uolentibus usus indiam nauigare contra uulturnum et septentrionem est desertum magnum usque ad terram sanctam per quod transitur in syriam non pauci

Explicacio de quibusdam opidis et notabilibus locis terre promissionis. ·B· Primo in Nazareth locus annunciacionis ostenditur. Et in capella ibi edificata erant tria altaria et intra capella excisa de petra in rupe sicut et locus noster et fontis et magna pars ciuitatis est erat excisa de rupe que nunc apparet. Ibi sinagoga in qua christus legit ysaiam [illegible]. Contra ciuitatem ad austrum ostenditur locus quod dicitur saltus domini ubi uoluerunt eum iudei precipitare [illegible] cecidit de manibus eorum et subito inuentus ostenditur in latere montis alio et uidentur ibi uestigia uestimentorum eius. Et sciendum quod ad locum annunciacionis descendi per gradus subterranea sicut et ad plura alia loca in terra ad opus est et nunc reedificata destructiones ecclesiarum. Sed ciuitas nunc reficitur loca in terra ad opus est. Mons tabor et hermon minor et hermonium uilla endor et nayn sicut habitant. Sub monte non ut mons tabor et hermon minor et hermonium uilla endor et nayn ficabant. Sub monte cedar gloriosa ciuitas et iezrael et fere tota latitudo campi magni esdrelon. ·C· cedar per litus aqlon contra maris Galilei hic munita in monte sanyr [illegible] transit uia que ducit achon per litus aqlon contra maris Galilei hic minuta in monte sanyr. Sueta dicitur uia maris ysaias ix et transit per sanctum Georgium. Cedar que iter in amman. Sueta unde fuit baldach suites. Sub hac ciuitate uersus cedar sunt facies de aram et syria et moab et amon et tota terra orientalis circa fontem fiale [illegible] pro loci amenitate toto estiuo tempore nundinas tenent et erigunt tentoria diuersorum colorum que de cedar aspiciunt tabernacula cedar. Dotain gratum prebent aspectum que in canticis uocantur tabernacula cedar. Dotun uersus aquilonem circa tapne et de toto est placens aspectus usque fasga et usque iericho. Dotaym. Equinoccium est ad opidum et ad uallem in qua ostenditur cisterna in qua posita fuit ioseph et est iuxta uiam que de Galaad ueniens conuergit in bethsayda cum uia que de syria ducit in egyptum ascendens de cotay iuxta montem betulie et in parte campi esdrelon procedens sub monte tabor ad sinistram per campum magedo ascendit montem effraym et per gazam ducit in egyptum. Ebron antiqua destructa est in qua regis

uit dauid. Sed ad ij iactus sagitte uersus meridiem declinando ad orientem est ebron noua ubi erat spelunca duplex. Ab ista spelunca ad occidentem ad iactum sagitte est campus damascenus. Ad australem autem de longa fodit ager ad iactum sagitte. Cham occidit abel sed ab ecclesie fossura ad orientem ad ij iactus sagitte in monte quodam ad latus ebron uetus est spelunca xxx pedum in longum et in latum ubi adam et eua luxerunt mortem abel. Tyberias In cuius australi parte sunt balnea medicinalia. In tyberiade et toto latere aquilonari maris galilei usque damascum est regio decapoleos sic dicta a x ciuitatibus principalibus que sunt tyberias, sephet, cedes, neptalim, asor, cesarea, capharnaum, Bethsayda, corozaym, bethsayda. Vocatur et decapolis Roboh. terra kabul. Galilea gentium. yturea. extenditur tamen yturea usque Beritum. Kabul saraceni uocant zabulon quod sonat displicencia de quo iij reg ix. Samaria hic situm placet proximum uersus mare a monte carmeli usque ioppen. Sychen sed neapolis iuxta ad ij iactus sagitte est fons iacob de quo io iiij. Ad cuius dextram est Garisi quem putant samaritanam et dixisse qui dixit patres nostri in monte hoc adorauerunt. Ad sinistram uero opidum quod putant antiquam syche. Et sunt syche putant opidum tebes et distant ab inuicem per ij iactus sagitte. Secundo quod fonti adiacet predium quod iacob dedit ioseph et est uallis longa fertilis et amena ualde. In sychen sunt ossa ioseph. Iezrael que et charetha in loco aliquantulum edito aspicit totam Galileam et usque ad montes Seyr et tabor et carmelum et montem effraym. De montibus. Montes a libano inchoantes usque inter desertum ubi fuerunt filij israel sunt unus mons continuus quamuis per uarias ciuitates habeat diuersa nomina. Mons tamen Galaad ceteris altior preminet iuxta illum se. Galaad tamen in capud libani et Antiliban cum libano protenditur ultra tripolim forte per v leucas et nusquam recedit a mari per ij leucas nisi iuxta tripolim a quo distat quasi per iij. Alicubi ita accedit ad mare ut nulla reliquatur uia. Inundat autem in opidis tirobus maris uiuus iuxta illud olee xiiij memoriale eius

61. Grid map H, drawn at Naples, 1329 or soon after; east at the top. The most carefully drawn of the surviving grid maps, with minutely detailed buildings marking the towns. Paris, Bibliothèque nationale de France, ms latin 4939, ff. 10v–11r.

symbol. Map B may well have been the earliest of the maps to be drawn; the sections of text that give the manuscript a date later than 1320, the date of map A, are additions that may be later than the map. It is worth noting that maps A and B are slightly smaller than maps C to G.

However, all seven Venice maps are alike in that none of them was produced with meticulous care and, indeed, one need not rule out some process of mass production. The grids certainly are carefully ruled, but not quite so precisely as the copying grid on the Archivio di Stato map. On every map we can find at least one red circle without a place-name, at least one place-name without a red circle. Some of the circles marking towns are brown instead of red[27] and we do not have to search hard to find places entered in the wrong square according to Sanudo's list.[28] Short rivers are apt to be omitted, while it is not easy to follow the tribal boundaries; some areas seem to be allocated to none of the tribes, occasionally some boundaries are drawn in brown instead of red[29] and there is occasional confusion between a boundary and a river.[30] All seven maps seem to have been constructed in the same order. The grid was drawn first, then on it were entered successively the outlines of mountains, the town symbols and place-names, the outlines of sea, lakes and rivers, and the further notes on the map, and it was probably now that the mountains and water were coloured; finally the boundaries and names of the tribes were added with the index letters to the notes below the map.

27 E.g. on map D.

28 Thus Gadara is in square 7.24 according to Sanudo's list (*Gesta dei per Francos*, ed. Bongars, II, p. 246; BL Add. MS 27376, f. 139v), but appears in square 7.23 on maps E and F. However, we must also take into account errors in the copies of Sanudo's list; the six manuscripts have never been collated. Bongars' edition in *Gesta dei per Francos* is from the text accompanying map C.

29 E.g. on map C.

30 E.g. on map D.

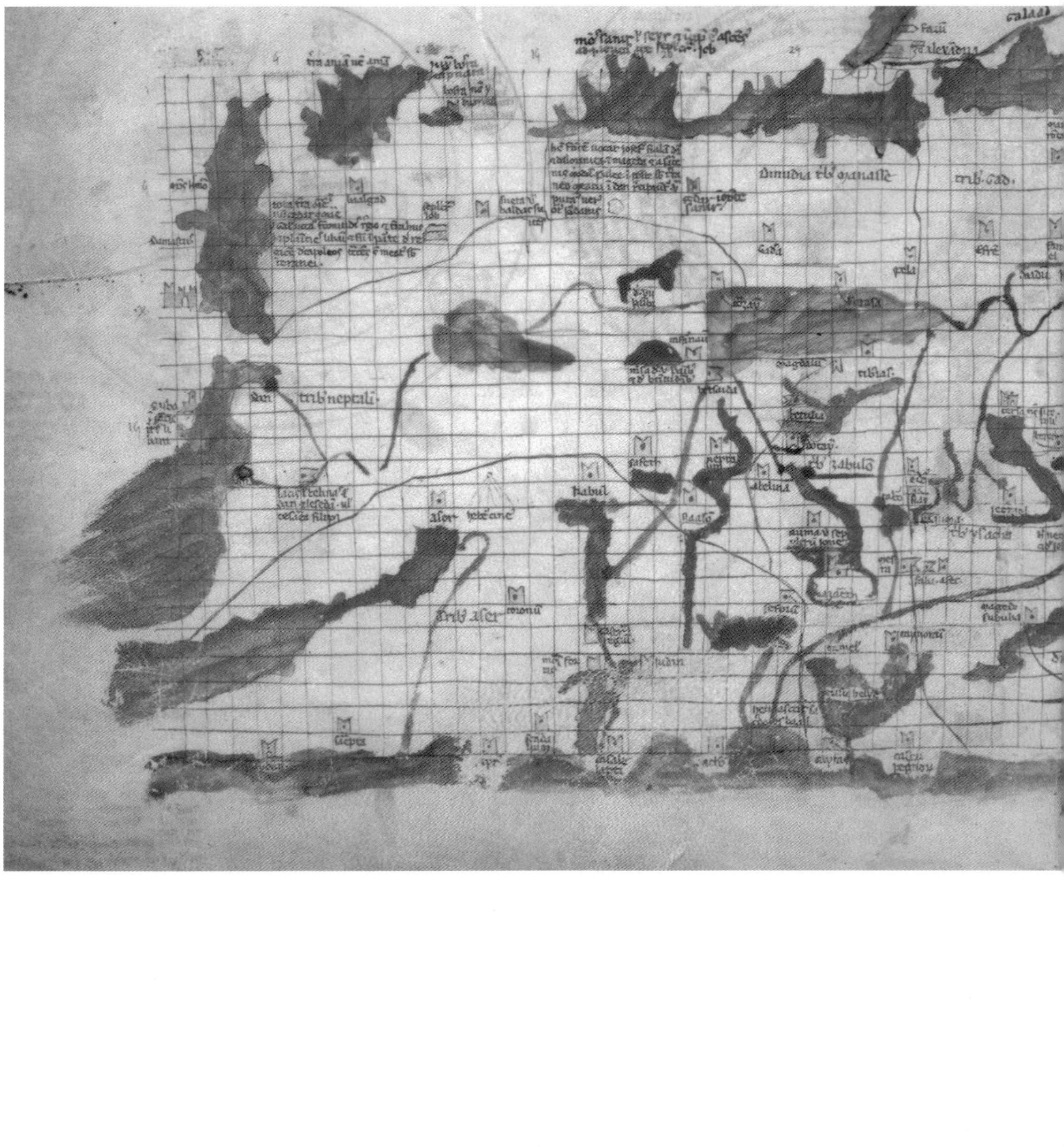

Running corrections appear here and there. Thus on map F the well of living waters and its two streams were drawn first at Acre, then erased and redrawn at Tyre, and we can see place-names that have been erased on map C, erased and rewritten on map G.[31] On other maps besides B there are later additions, among them several towns and sites on map A,[32] again entered with a finer pen that the rest, and on maps F and G the names of Sodom, Gomorrah and Lot's wife (*uxor loth*). On the left-hand page of map D we can see the reversed outlines of the Red Sea and the Persian Gulf, part of the map of the eastern Mediterranean with lands to the east drawn on the other side of the leaf. More confusingly, as a curator noted in 1934,[33] at some time in the course of conservation some of the leaves in the manuscript containing map E had been placed upon each other while still damp, so that ink was transferred from one to another; the left-hand page of the map of the eastern Mediterranean had been placed on the right-hand page of map E (they are two pages of the same bifolium) and had transferred to it not only several place-names in reverse but also radiating rhumb-lines, which, since they appear on the sea, one might easily suppose belong to map E itself.

Turning to the maps drawn at Naples, map H is unquestionably the finest of all nine grid maps. Carefully and neatly drawn, it marks towns not with a red circle or simple symbol, but with tiny pictures of towered buildings of differing designs, some of them fantastic; they are shaded yellow, while a splash of red highlights the initial of

31 *Herodium*(?) on map C; *Betsaida*, now *Betsayda*, on map G.

32 E.g Deborah's palm and Rachel's tomb.

33 The note, on the flyleaf, has the initials HIB, i.e. H.I. Bell.

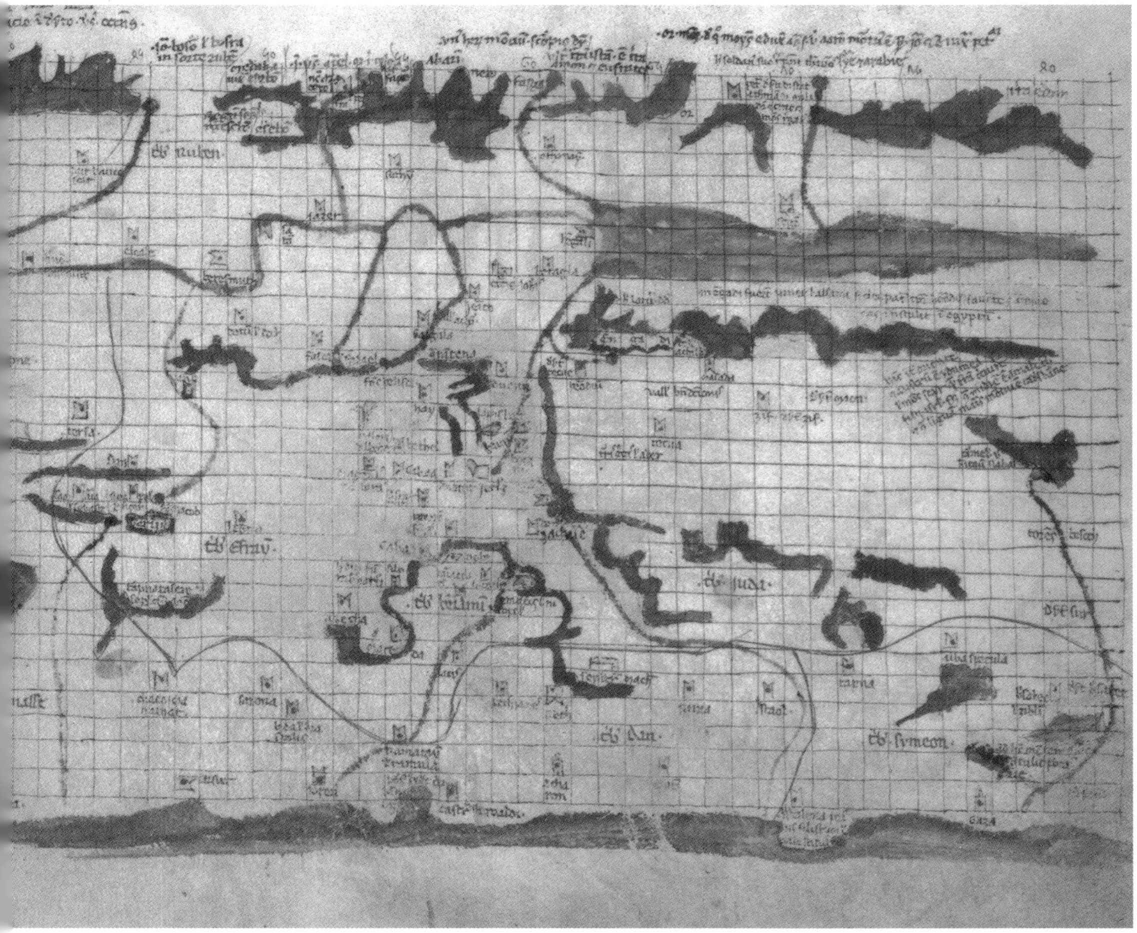

62. Grid map I, drawn at Naples, 1334 × 1339; east at the top. On the upper half of the same opening is a map of the Near East, partly reproduced in **[21]**. Vatican City, Biblioteca Apostolica Vaticana, ms Vat. lat. 1960, ff. 265v-266r.

many of their names. At Damascus there is quite an elaborate picture of town walls and citadel, with green trees. The grid lines are ruled alternately in red and brown ink, and the tribal boundaries are in gold, with none of the confusions of the Venice maps – though one, drawn in black between the tribes of Dan and Symeon (bottom right), must have been added later to correct an omission. Mountains are carefully coloured with dark brown, yellow-brown and grey shading, while the sea and lakes are coloured not with a simple wash but with drawn lines of waves. The map is in keeping with the rest of the manuscript; it was clearly a presentation volume, but we do not know to whom it was given. It was annotated by Giovanni Boccaccio, who drew on Paolino's work for his own historical encyclopaedia.[34]

Map H seems to have been drawn in at least broadly the same sequence as maps A to G (though map J, we shall see, may point to a rather less systematic process). Certainly, however, the grid was set out first, and the gold lines of the tribal boundaries were drawn last, overlaying any features they crossed. There are a few running corrections: town symbols at the end of strips 22 to 24 were originally drawn one line too low and were redrawn, while the lines of the rivers south of Jerusalem have been adjusted. Rough outlines of buildings at the north-east end of the sea of Galilee and, drawn sideways, at the north end of the Dead Sea are clearly later additions. More puzzling are changes to the course of the Jordan and its first tributary from the east to the south of the sea of Galilee; originally drawn in just the same form as on the other grid maps, they have been erased and redrawn in a not very different pattern – as a running correction to judge from the town symbols and wording which reveal no second hand at work **[64]**. The changes are a little hard to assess: not only are they in or beside the gutter in the middle of the opening, but the curving of the parchment seems to have caused the rivers' colouring to flake off.

34 Giuseppe Billanovich, 'Autografi del Boccaccio nelle Biblioteca Nazionale di Parigi (Parigini lat. 4939 e 6802)', *Atti della Accademia Nazionale dei Lincei: rendiconti, classe di scienzi morali, storiche e filologiche*, 8th series, 7 (1952), pp. 376-82.

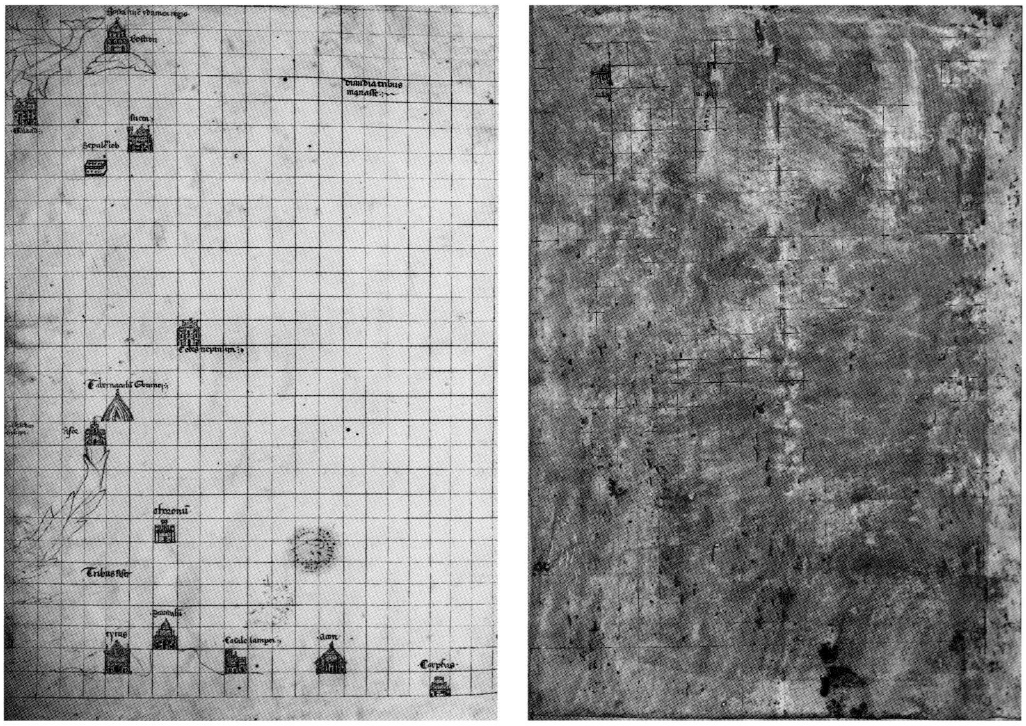

63. Grid map J, drawn at Naples (?), 14th century; east at the top. Apparently the start of what would have been a large and extremely fine copy of the grid map; for some reason it was abandoned and the parchment used in binding another manuscript. Paris, Bibliothèque nationale de France, ms latin 15363, ff. Iv-IIr.

Beside map H we should place map J, another copy of the grid map that is not unfinished so much as scarcely begun. The parchment, cropped at each side, has survived because it was used as a pastedown and flyleaf in binding a fourteenth-century copy of Pierre Auriol's commentary on the *Sentences* of Pierre Lombard. The right-hand page that was pasted to the board has suffered badly, but we can still see the symbols and names of Jaffa and Arsuf. On the left-hand page, the verso of the flyleaf, are parts or all of the names and symbols of seventeen towns or other sites, the outlines of some mountains and the names in red of two tribes, Asser and the half-tribe of Manasseh beyond Jordan. Had the map been completed it would have

64. Running corrections on grid map H. Some rivers have been erased and redrawn on both sides of the central gutter of the opening. Paris, Bibliothèque nationale de France, ms latin 4939, ff. 10v-11r.

outshone map H; it is drawn on a larger scale and the buildings that mark places are more carefully elaborated, with some fine details in red. Apart from the tomb of Job and the tent of Heber the Kenite (*Tabernaculum Eburnei*)[35] the form of each symbol is unrelated to that on map H. Why the map was abandoned and its parchment used as scrap we cannot tell; Baalgad was misnamed *Galaad*, but this was hardly an irreparable mistake. We may suppose that a complete map in this form was actually drawn but is now lost – but this can be only a guess.

Map I, probably the latest of all ten maps, is the only one to have pinpricks in the parchment for ruling the grid – along top and bottom but not at the sides. Like maps C to G, but unlike map H, it is in a separate quire of maps, here appended to the two texts, the *Great chronology* and the *Satiric history*, and it occupies the lower half of an opening that it shares with the map of the eastern Mediterranean with lands to the east. Apart from Jerusalem,[36] it marks towns with a two-pointed tower of mitre-like shape, a symbol that recurs in the manuscript's other maps; like map H the town symbols are simply coloured yellow and some of the place-names have a touch of red on the initial letter, but unlike map H each town symbol has a red dot in the centre. The grid was drawn first and the town symbols and mountain outlines were added as a single operation.[37] The few outlines of water areas were then drawn, place-names and notes were written and the mountains and water were coloured. The red of the tribes' names and boundaries and on the town symbols and names was added last.[38] The overall result is more akin to the briskly produced maps A to G than to the carefully crafted maps H and J.

We have no more of these grid maps, though their outlines reappear in many later maps without the grid.[39] From the fifteenth century, however, we have descriptions of two maps of Palestine with this same grid used to locate places, though in neither case does the map itself survive. Both texts name far fewer places than Sanudo and though many are the same – and in the same squares – as in Sanudo's list, both name others that he does not. While the two texts are quite different, there is an odd echo between them in that both, unlike Sanudo, specifically mention the cities of refuge, the places of sanctuary named in the book of Joshua – as we have seen on the ninth-century map of the Land of Promise (chapter 4, pp. 21-2) – for anyone who killed unintentionally.

What is probably the earlier of the two descriptions is in an account of the Holy Land by John Poloner, who was there in 1421-2; he was probably German, but we know nothing else about him.[40] Towards the end of his account, after descriptions of Jerusalem and a series of journeys to other cities, is a section headed 'Division of the Holy Land' (*Divisio Terrae Sanctae*) and there he names and describes successive places, identifying the square where each would be found on the grid.[41] However, he does this much less systematically than Sanudo, with much interspersed information, and unlike Sanudo he offers no explanation of the grid or of how it works; he must have assumed that with the map before us no explanation would be needed. However, he explains other features of the map and in using the first person singular he shows that he drew it himself or – at very least – altered an existing map. To distinguish the kingdom of Og, king of Bashan, from the rest he had painted it yellow (*citrinum pinxi*) and the neighbouring region of Bethany white (*album feci*), while the land of Moab

35 Correctly *Tabernaculum Eber Cinei* (Judges 4.11-21).

36 Jerusalem is marked with a strange symbol that might represent the front and one side wall of a square walled town.

37 Thus the hill outline breaks for the town symbol at Bosra, symbol and hill are integrated at Bethulia, the symbol breaks for the hill at Caymon (*Caynmaim*).

38 The red dot in each town symbol was added after, not before, the symbol was coloured yellow.

39 A map of 1350 by Domenico Pizigano, now lost to sight, with names specifically from Marino Sanudo, 'differs considerably' from the grid maps and may have been a version of the map of the eastern Mediterranean (G. Libri, *Monuments inédits* (London, 1862, re-issued 1864), plate 58; sale catalogue, S. Leigh Sotheby and John Wilkinson, *Catalogue of the Libri collection, 25 July 1862*, pp. 131-2). I owe these references to the kindness of Dr Chet Van Duzer.

40 The text is edited in *Descriptiones*, ed. Tobler, pp. 225-81, with discussion and notes on pp. 497-522; an English translation by Aubrey Stewart was published as *John Poloner's description of the Holy Land (circa 1421 A.D.)*, Library of the Palestine Pilgrims' Text Society 6 (London, 1897).

41 *Descriptiones*, ed. Tobler, pp. 255-63. There are other casual references to locations on the grid on pp. 254, 268.

and Ammon was again yellow. He had marked St George's church at Lydda with the criss-crossed oval of a tortoise's shell (*signavi cum arcu testudinis*)[42] and he mentions other signs on his map that are not on the grid maps: three stars to mark the cities of refuge (all six are named), a V with one dot and reversed with two at the places where Christ fed the five thousand and the four thousand and a circle linked with a semicircle at Caesarea Palestina.[43] Seven manuscript copies of Poloner's text are known, but none is accompanied by a copy of the map.[44]

We know the other description from only one manuscript, in which it is the first of a little collection of fragments of text describing the Holy Land.[45] Its heading probably applies to this text alone, not to the whole collection – 'Explanation of the map of the Holy Land' – and below this another hand has added 'by Master Maurice Paris'. Of him we know only what he himself tells us, that he painted (*pinxi*) the map (*figura*) on Mount Sion.[46] The other texts in the collection are extracts from Burchard's work, apart from one on the Dead Sea based on Rorgo Fretellus and a description of Jerusalem that is at least related to William of Tyre's. Again there is no explanation of the grid, but the text's list of places on the map is more business-like than Poloner's, rather more like Sanudo's, and much comes from Sanudo's list – the account of the Sueta summer fair, for instance, and the description of Bethsaida, home of Peter, Andrew and Philip, beside the road from Syria to Egypt, where the Galilee shoreline curves to the south. On the other hand it includes several places that Sanudo does not: Madaba, Ramoth-Gilead and others. Maurice, or more likely a copyist, shows signs of boredom when he reaches the seventh row, saying that it includes further places that can be seen on the map; he leaves out rows thirteen to sixteen without comment, and at the end of row eighteen gives up altogether, effectively telling his readers to look at the map for themselves. A note at the end explains that the green lines are watercourses, the red lines roads. The text has never been published and it is printed here as an appendix to this chapter, pp. 126-7.

The grid maps have long been known to historical scholarship. An engraving of map C was published in 1611 in what is still the only edition of Sanudo's *Book of secrets*,[47] and since then there have been many reproductions of one or other of the maps, latterly often in colour.[48] They were of particular interest to historians looking for early steps towards 'scientific' – measurement-based – cartography, as they might be seen as a precocious anticipation of the grid of geographical co-ordinates, 'a sort of unscientific substitute for lines of latitude and longitude' as C.R. Beazley put it in 1906.[49] In the 1890s several scholars had advanced our knowledge of the maps: Konrad Kretschmer identified Vesconte as the original mapmaker, Reinhold Röhricht and Cornelio Desimoni showed that they are related to the Archivio di Stato map, though raising needless difficulties in not appreciating that this is itself an imperfect copy, while Arturo Magnocavallo listed the manuscripts of Sanudo's work as a first step, never followed up, towards a new edition.[50] A new edition was again mooted in 1976, when Franco Cardini also published a list of the manuscripts with other introductory work.[51] However, looking at the maps individually rather than as a group has become possible only since 1973, when Bernhard Degenhart and Annegrit Schmitt published their masterly analysis of the artwork in the manuscripts of Sanudo's and Paolino's texts: they saw the maps and plans as a part – if a separate

42 However, *snail* is a not impossible translation of *testudo* (cf. R.E. Latham, *Revised medieval Latin word-list from British and Irish sources* (London, 1965), p. 482) and the spiral of a snail's shell might seem more likely here.

43 *Descriptiones*, ed. Tobler, pp. 257-8, 261-2.

44 Reinhold Röhricht, *Bibliotheca geographica Palaestinae*, ed. by David H.K. Amiran (Jerusalem, 1963), pp. 106-7; Tobler drew on two manuscripts for his edition (*Descriptiones* ed. Tobler, pp. 520-1).

45 Munich, Bayerische Staatsbibliothek, Clm 18736, ff. 201r-211r.

46 f. 203v.

47 *Gesta dei per Francos*, ed. Bongars, II, plate II (between pp. 282, 285)

48 E.g. map G in Joachim Lelewel, *Géographie du Moyen Âge: atlas* (Brussels, 1849); map C in Nebenzahl, *Maps of the Holy Land*, pp. 44-5; map E in Harvey, *Medieval maps*, p. 79; map H in *Couleurs de la terre: des mappemondes médiévales aux images satellitales*, ed. by Monique Pelletier and Catherine Hofmann (Paris, 1998), p. 39.

49 Beazley, *Dawn*, III, p. 529.

50 Kretschmer, 'Marino Sanudo'; Röhricht, 'Karten und Pläne I', pp. 9-10; Röhricht, 'Marino Sanudo', pp. 84-100, plate 2; Desimoni, 'Carta della Terra Santa', pp. 243-53; Arturo Magnocavallo, 'I codici del "Libro secretorum fidelium crucis" di Marin Sanudo il vecchio', *Reale Istituto Lombardo di Scienze e Lettere: rendiconti*, 2nd series, 31 (1898), pp. 1113-27.

51 Cardini, 'Per un' edizione critica', pp. 191-3.

part – of this artwork, and illustrated and discussed them.[52] Their work has opened a new stage in our understanding of the grid maps.

APPENDIX

EXPLANATION OF THE MAP OF THE HOLY LAND BY MAURICE PARIS

Munich, Bayerische Staatsbibliothek, Clm 18736, ff. 201r–204v

Expansions of abbreviations are in italics. Full stops are editorial additions except the first two in the second paragraph, where they correspond to pen-strokes in the manuscript.

[f. 201r] Decl*ar*acio mappe t*er*re s*an*cte ⌜p*er* m*a*gist*ru*m Mauriciu*m* P*ar*isiense⌝[53]

In p*ri*mo eni*m* quadro quod e*st* v*ersus* oriens incipiendo ab aquilone et p*ro*cedendo v*ersus* austru*m* in quadro ⌜41⌝[54] est awr q*ue* metropols dicit*ur*. In ⌜54⌝[55] q*ua*dro e*st* petra des*er*ti que nu*n*c mons regalis dicit*ur* vbi soldan*us* suos repo*n*it thesauros arabie et egipti ⌜ecia*m* i*n* castro m*ar*moreo⌝[56]. In 2° a*utem* spacio et 12° quadro est bostra et regio bostron dicit*ur* vel ydumea vn*de* ysaie 62. In 3° spacio quadro v*er*o 48 est bosor. In q*ui*nto a*utem* quadro ei*us*de*m* spacij locatur esebon. In 52 modoba. In 61 e*st* corroaym sicut p*resen*s docet fig*ur*a no*n* vt mappa mundi q*ue* fallax e*st* in his p*ro*ui*n*ciar*um* et i*n*sular*um*. In q*ua*rto itaq*ue* spacio et q*ua*dro 23 [f. 201v] est golon ciuitas fugitiuoru*m*. In 36 est masceronca vbi latuit dauid et b*eatu*s ih*o*ha*nn*es baptista po*sit*us fuit i*n* carcere. In 44° e*st* yair vel lamoson alias Amochseir. In 50 quadro p*er* dimidiu*m* miliar*e* ab eberon cont*ra* occidens est dibon sub fibria montiu*m* arabie. In 53 e*st* saliama. In 55 est serijn. In 58 e*st* noryn. In q*ui*nto quadro 8° spacio est baldach. In 21 e*st* cadar ci*ui*tas munita i*n* monte san*n*ij *u*el sanijn vnde e*st* via q*ue* ducit ad accon iuxta litt*us* aquilonale maris galilee et p*er* cedar maram q*ua* damascu*m* vn*de* in ysaia di*citur* via maris quia o*mn*i*no* p*ro*cedit in littore maris galilee[57] transiens per [f. 202r] vallem t*er*re aser a sinistris h*ab*ens ciuitate*m* sereth et vna p*ar*s dicit*ur* galilea ge*n*tiu*m* que ibide*m* in iordane tra*hi*tur. In 31 quadro est ramothgalaad. In sexto spacio et 13 quadro est sueta vn*de* fuit baldach suites. Sub hac ciuitate v*ersus* cedar soliti sunt sarraceni *con*uenire de ara*bi*a et mesopotamia et amon et syria moab et ex tota illa t*er*ra orientali circa fonte*m* ibidem p*ro*pter loci ame*n*itate*m* toto estiuo *tem*po*re* nundinas celebrant et erigu*n*t tentoria diu*er*sor*um* color*um* q*ue* de ciuitate cedar aspicie*n*tib*us* pulcru*m* p*re*be*n*t asp*ec*tum vn*de* ca*n*ticor*um* P*ri*mo Sicut tab*er*nac*ul*um cedar. In 7° spacio et q*ua*dro vndeci*m*o [f. 202v] est sepulc*ru*m Job. Deinde cont*ra* austru*m* p*er* vna*m* leuca*m* incipit ascens*us* montis sanijr. Cet*er*a loca ei*us*de*m* quadri clare*n*t i*n* fig*ur*a. In 8° spacio et quadro 19 est loc*us* vbi d*omi*n*us* pauit q*ui*nq*ue* milia ho*m*inum. In 61 est becagla vbi filij is*ra*el pla*n*xeru*n*t Jacob p*at*rem suu*m* q*uan*do tuleru*n*t eu*m* de egipto et distat p*er* vna*m* leucam a iordane. In nono et xxij quadro e*st* corrosaim in p*rinci*p*io* ma*ri*s galilee de qua mathei xi° Ve *ti*bi corrosaim. In 27 est garasa de qua gerasinor*um* r*e*gio. In vndeci*m*o spacio quadro 2° diuidit*ur* liban*us* a mo*n*te hermon sub qua e*st* damasc*us* a parte septe*n*trionali et in hac diuisio*n*e montiu*m* e*st* via qua it*ur* i*n* emath. [f. 203r] In 12° spacio et xxij quadro est bethsaida ciuitas b*ea*ti petri andree et philippi iux*ta* viam que ducit de syria i*n* egiptu*m* vbi mare curvar*e* incipit v*ersus* austru*m* vnde Jho*ha*n*n*is p*ri*mo. In vicesi*m*o est tyberias a qua mare tiberiadis dicit*ur*. In 17° spacio et quadro 6° e*st* lacis quem ceperu*n*t filij dan in nomine

52 Degenhart and Schmitt, 'Sanudo und Veneto', pp. 60–87, figs 143–64.

53 Added in a different hand.

54 Added in margin; '55' struck out in text.

55 Added in margin; '68' struck out in text.

56 Interlineated in a different hand; 'castro marmoreo', added in margin, struck out.

57 'et per cedar mara' struck out.

patris sui et vocauerunt eum lesdan que et cesarea philippi appellatur. In 18 est castrum kabul quod zabulon dicitur de quo 3° regum capitulo 5°. In 43 est sebaste vel samaria[58] cuius situs pulcher erat valde et latus versus mare a monte carmeli versus joppen. In 45 est sichen nunc neapolis ad duos iactus sagitte ⌜est⌝[59] fons Jacob de quo Jhohannis quarto discemus. [f. 203v] Ad cuius dextram est mons gerasin quem mulier samaritana domino ostendit quando dixit In monte hoc patres nostri adorauerunt. A sinistris est opidum antiquum sichen et secundam et distant ab inuicem per duos iactus sagitte.[60] In 54 est gaba a qua saul fuit de quo regum primo. In 18[61] et quadro vndecimo est assur munita ciuitas que omnia clarent in figura presenti quam pinxi in sancto monte syon *etcetera*.

Ista est longitudo terre sancte siue promissionis 83 leucarum siue 42 longorum miliarium siue teutonicalium ducentorum 20 ytalicorum et protenditur a dan que est sub monte libano ab aquilone [f. 204r] contra austrum vsque barsabee que est prope desertum egipti. Latitudo vero illius terre sancte est a mari occidentali siue meditarraneo ab occidente et protenditur versus orientem vsque ad fimbriam montium arabie per 28 leucas seu 14 miliaria teutonicalia siue magna et 70 ytalica quia vna leuca continet duo miliaria[62] maritima et duo leuce secundum hanc figuram faciunt vnum miliare almanicum. Tota itaque terra sancta in 83 quadra protrahitur et in 28 spacia secundum latitudinem tractis lineis super priores lineas transuersaliter ab oriente in occidens vel equo. Et per latum ponitur spacium scilicet per lineas rubeas a sursum versus deorsum protractas. Per longum vero scilicet per lineas nigras [f. 204v] A septentrioni versus meridiem protractas. Vocatur quadrum et spacium siue quadrum vocatur leuca et continet duo miliaria maritima et est medium miliare almanicum.

Nota. Per tractulos virides huius figure riuuli fontes et flumina figurantur. Per tractulos autem rubeos meatus viarum designantur.

58 'sam' added in margin.

59 MS 'et'.

60 Cf., in Sanudo's list, 'ad sinistram ver' oppidum, quod putant antiquum Sicen; et secundam Sichen putant oppidum Thebes; et distant ab invicem per duos iactus sagittae' (*Gesta dei per Francos*, ed. Bongars, II, p. 247).

61 'est' struck out.

62 'ytalica' struck out.

11 THE LATER LARGE BURCHARD MAPS
14th to early 15th century

A New York, Pierpont Morgan Library, MS M.877 **[65]**. Parchment, 52 × 165.5 cm; two membranes, glued together, with 2 cm overlap, width of each (left to right) 73 cm, 94.5 cm. Grey-brown and red ink, with purple-brown, green, olive-green and red paint. Dorse not accessible.

B Bruges, Archives de l'état, Chartes avec numéro bleu 11458 **[68, 69]**. Parchment, two fragments, each approximately 32 × 20 cm. Brown and red ink. On the dorse of each, parts of a list of records, 17th century (?).[1]

C Oxford, Bodleian Library, MS Douce 389 **[66]**. Parchment,[2] 40.5 × 215 cm; three membranes, glued together, with 15-20 mm overlap, total width of each (left to right), 75 cm, 69 cm, 74 cm. Brown, blue and red ink, with blue, dark brown, light brown, dark red-brown, green, olive-green, pink and red paint. Dorse not accessible.

THROUGH ITS ABRIDGEMENT in the grid maps, the influence of the large Burchard map extended far into the age of printing. In the fifteenth century successive derivatives of the grid maps, first in manuscript, then as woodcuts, appeared among the new maps that appeared in late manuscripts and printed editions of Ptolemy's atlas and elsewhere;[3] so many copies of the original grid maps must then have existed that it is not surprising that they were drawn on in this way. While these derivatives have no grid, the 83 spaces from north to south, 28 from east to west, are duly marked in the margins. Subsequently we can trace the grid maps' outline in maps of Palestine down to the seventeenth century.[4] The unabridged large Burchard map did not achieve the same longevity, but we have three copies from the fourteenth or early fifteenth century and a contemporary account of a fourth and it is likely that many more were made, for the survival rate of large sheet maps from any age is bound to be extremely low. Indeed, while two of the three are complete maps, the third (here map B) was cut up as scrap parchment and we know it only from two pieces that were used as covers for archives. Complete or fragmentary, or only described, all are of great interest.

In 1956 the Pierpont Morgan Library's Fellows, with the particular help of two donors, bought map A from the New York dealer H.P. Kraus as a gift to the Library. Kraus had bought it the previous year from an unidentified Swiss dealer and we may reasonably suppose that it had an Italian provenance; its hand and style show that it was made in Italy in the first half of the fourteenth century. A short descriptive note appeared in the Librarian's annual report to the Fellows in 1957, but it has otherwise been noticed only in summary catalogue entries.[5] No detailed account of the map has been published, nor any workable reproduction.

1 These are printed by Van de Walle, 'Deux fragments', p. 56n.

2 Not paper, as *Itineraries of William Wey*, p. xvi.

3 Thus, BL Add. MS 15760, ff. 58v-59r (Henricus Martellus Germanus, Book of Islands, *c.* 1470); Vatican City, Biblioteca Apostolica Vaticana, ms Vat. lat. 3811, ff. 45v-46r (Ptolemy, Geographia, 1474), reproduced in Nebenzahl, *Maps of the Holy Land*, pp. 58-9; editions of Ptolemy reproduced in *Claudius Ptolemaeus Cosmographia: Ulm 1482*, ed. by R.A. Skelton, Theatrum orbis terrarum, 1st series 2 (Amsterdam, 1963), and *Claudius Ptolemaeus Geographia: Strasbourg, 1513*, ed. by R.A. Skelton, Theatrum orbis terrarum, 2nd series 4 (Amsterdam, 1966).

4 E.g. Catherine Delano-Smith and Elizabeth Morley Ingram, *Maps in Bibles 1500-1600: an illustrated catalogue*, Travaux d'humanisme et Renaissance 256 (Geneva 1991), pp. 59-60, 67-8; Nebenzahl, *Maps of the Holy Land*, pp. 94-7, 110-13, 118-19.

5 Adams, *Seventh annual report*; C.U. Faye and W.H. Bond, *Supplement to the census of medieval and Renaissance manuscripts in the United States and Canada* (New York, 1962), p. 368; *The Pierpont Morgan Library: a review of acquisitions 1949-1968* (New York, 1979), p. 5. Surprisingly, there is no notice of it in *Major acquisitions of the Pierpont Morgan Library 1924-1974: mediaeval and Renaissance manuscripts* (New York, 1974).

65. The Pierpont Morgan map, first half of the 14th century; east at the top. This is very like our earliest large Burchard map **[48]**, but is more ornate, more carefully drawn and in better condition. The retention of the parchment neck at the right-hand end shows that it was meant to be kept rolled up. New York, Pierpont Morgan Library, MS M.877.

66. The Douce map, late 14th century; east at the top. While a single hand wrote all the inscriptions, each of the map's three sheets of parchment was the work of a different artist. Oxford, Bodleian Library, MS Douce 389.

In its overall size and general proportions the Pierpont Morgan map is very like the Archivio di Stato map. There is, however, no list of places at its north end and in the south the map terminates some 20 cm before the end – the neck end – of the parchment. The Librarian's report in 1957 suggests, convincingly, that the map 'was intended to be rolled up from the left, the neck flap having been allowed to remain on the right-hand skin for use in gripping and docketing'.[6] However, the map must long have been kept in a frame, for on a strip of from one to two centimetres along the other three sides the parchment is distinctly paler and there is a series (perhaps two series) of nail holes. As the map's detail stops short of this strip we may wonder whether the mapmaker did not have framing in mind when he drew the map. The map now, with no part of the parchment masked, is attached to a mount set in a glazed frame, so its dorse cannot be seen.

The whole of the map's surface is slightly rubbed, most strongly at the south end, where some paint has flaked off, and a strip along the centre of the left-hand membrane seems to have become wet, resulting in discolouring. There is some damage to the edge of the parchment and at the lower right end damaged parts have been trimmed away; originally the neck was probably the same shape below as above, so that the southernmost river, the Brook Besor, extended all the way to the Mediterranean coast. Overall, though, the map's surface is in good condition and its details are clear and bright. It is in a significantly better state than the Archivio di Stato map.

It was also drawn with much greater care and is a very handsome map. It uses few colours, but to great effect. All water is green, the mountains and the hillocks below the town symbols are a purplish shade of brown and the tribes' names and boundaries, the wording on the lakes and a couple of regional names, as the plain of Moab (*Campestris Moab*), are all in red, while a splash of red highlights the initial letter of most of the longer inscriptions. As on the earlier map, each place has the picture of a building, drawn in a great variety of forms, many owing more to fantasy than to architectural reality; each is finely detailed with its details picked out in red and in a deep olive-green that differs from the bright green of the water. The shapes of the buildings are mostly unrelated to those on the Archivio di Stato map or on the finely drawn grid map H (chapter 10), though a few clearly traditional forms recur, such as the spire of Job's tomb and the triangular tent of Heber the Kenite.

In its content the map is closely related to the Archivio di Stato map; both must derive, perhaps at several removes, from the same copy of the large Burchard map, a copy that differed from that used for the grid maps. Thus the tribes of Reuben and Gad are misplaced in the same way and we see the same forms of some place-names; as on the earlier map we have *Gorsa* and *Dora* for the grid maps' *Gerasa* and *Arsur* and where the earlier map garbled Gath as *Thet*, Caymon as *Chaimborat*, they here appear as *Teth* and *Caynnobac*. Most tellingly, Jerusalem is again marked not with a symbolic building but with a cross within a circle, here conspicuously drawn with the cross red and the circle green and with its name – omitted from the earlier map – written beside it in red. On the other hand the names of Bosra, completely garbled on the earlier map, appear correctly here; the error was perhaps introduced by the earlier map's copyist.

6 Adams, *Seventh annual report*, p. 16. The Library's unpublished catalogue description notes that under ultra-violet light traces of indecipherable writing appear on the neck end of the parchment.

A very few places are shown that do not appear on the Archivio di Stato map, among them the house of Zacharias; they all appear on the grid maps and were probably left off the earlier large map for want of space or simply by accident.[7] More differences appear in other inscriptions. Some are identical on both maps or differ only in detail; thus the river Jor, *Secundus fons Iordanis* on the earlier map, is here *Secundus de fontibus Iordanis*. Some, though, paraphrase or elaborate those on the earlier map; thus by Megiddo,

Hic iosias occiditur a rege egipti

is here

Hic fuit occisus iosias a pharaone tiecas rege egipti.[8]

Where there are differences of this sort it is on the Pierpont Morgan map that the inscriptions are longer, sometimes adding further detail; possibly they are closer to those on the map from which both stemmed, abridged on the Archivio di Stato map. The long inscription that on the earlier map appears in the sea, west of Caesarea Palestina, is here placed on land to the east, again perhaps reflecting a more spacious layout.

On the map is the same grid as on the grid maps, 83 squares from north to south, 28 from east to west. It has been ruled carefully and accurately in ink: any ten squares consistently measure from 15.0 to 15.5 cm vertically and from 15.5 to 16.0 cm horizontally. The first part of the map to be drawn, it was presumably used as a copying device, but it seems odd that it was drawn so conspicuously and retained on the finished map. A possible explanation is what was suggested for the grid on the map in the work by Paolino Veneto who, unlike Marino Sanudo, did not make use of it as a guide to finding places on the map: Sanudo's book, with its map, had been so widely distributed that people had become accustomed to the idea that a map of Palestine might properly be marked in squares.[9] Whether or not this is the explanation, we can reasonably assume that the map cannot be earlier than about 1320, when the grid of 83 by 28 squares was devised as the basis of Sanudo's index, and that it was probably not drawn before the grid map had become widely distributed in the 1330s.

It is more puzzling that neither towns nor physical features occupy the same positions as on the grid maps. The town symbols, while broadly in the same areas, do not coincide at all exactly with the red circles on the grid maps or with Sanudo's list **[67]**. Even the circle and cross marking the key site, Jerusalem, in square 14.59 on the grid maps, are here in square 16.62. In the physical features, much simplified on the grid maps, there is less divergence, but mostly no exact correspondence. Thus, while the springs of living waters near Tyre are in squares 14.25 and 14.26 on both, the east shore of the sea of Galilee is in rows 9 and 10 on the grid maps but here in row 8, and the river that enters the Dead Sea from the east, in column 71 or 72 on the grid maps, is here in column 75. We have seen that the present map, along with the Archivio di Stato map, stems from one exemplar, the grid maps from another (and better) one, and the differing locations presumably reflect this, but how they came about we can only guess.

Once the grid had been ruled, the outlines of the town symbols were the first feature entered; many of them more or less follow the lines of the grid. Next, the outlines

7 Others are Elealeh, Dothan and Upper Beth-horon (*Bethron superior*; the grid maps have Lower Beth-horon, *Betoron inferior*).

8 II Kings (IV Kings in the Vulgate) 23.29, where the Pharoah's name is Nechao.

9 Above, pp. 116-17.

of coast, lakes and rivers were drawn; we see how at Jericho the river breaks for the symbol but not for the hillock below, clearly drawn later. The outlines of the hillocks and of the mountains were drawn next and then either the grey-brown names and other inscriptions were written in, or else the colours were applied to water, mountains, hillocks and town symbols – green first, then purple-brown, then red and olive-green to judge from the overlaps. Except on the lakes, all the writing was entered first in what is now grey-brown; the tribal and other names in red simply repeated, above or below, what had already been written in the grey-brown that was then erased on the left side of the map but retained on the right side. Along with the red wording, the tribal boundaries and red highlights on initials were added as the last stage of making the map. The outline drawings of Lot's wife and of Abraham with three angels at the tree of Mamre are later additions, in ink of a slightly different colour, and a much later hand has added 'Aque de Marun(?)' on either side of Kedesh-naphtali (*Cedar Neptalim*).

In 1928 Baudouin van de Walle described two fragments of a medieval map of Palestine, with a half-tone reproduction and line drawing of each.[10] They had come to light in the Archives de l'état at Bruges, where, folded vertically, they had served as covers for groups of seventeenth-century documents; the documents were listed on the outside of each cover, that is, on the back of each fragment of map, but it is not known to what archive they belonged. From the handwriting a colleague at the

10 Van de Walle, 'Deux fragments', pp. 56-75.

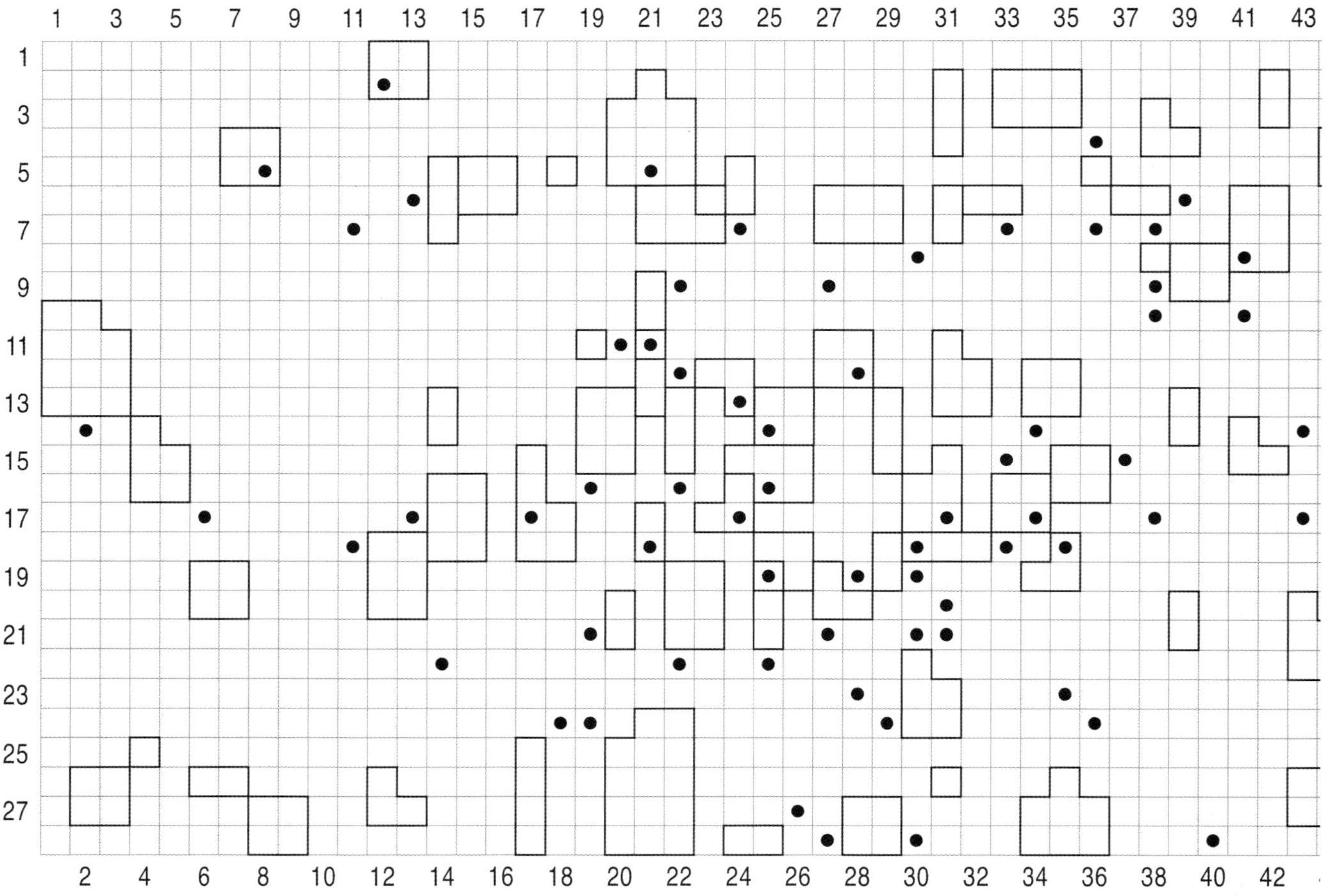

Archives du royaume at Brussels had suggested, reasonably enough, that the map had been drawn in Italy in the fourteenth century. East was at the top, and one fragment showed the area from the west side of the sea of Galilee almost to the Mediterranean coast; the other, the southern half of the sea of Galilee with the area to the east as far as Ramoth-Gilead and Macheronta, both somewhat displaced.

Though they are of considerable interest, there has been no further account of the fragments since 1928. Unfortunately, through a misplacement in the repository they cannot currently be made available, but there is no reason to suppose they are permanently lost. Luckily there are excellent black-and-white photographs and digital images of both fragments.[11] Van de Walle showed that the map is closely related to the Archivio di Stato map, and it is indeed a later copy of the large Burchard map. It must have been larger than any other version that we have; it measured about 65 cm from east to west, so we would expect its length from north to south to have been about 2.1 metres. However, its proportions may have differed from the other versions of the map that we know; comparing it with the Archivio di Stato map, Van de Walle pointed out that the distance from the sea of Galilee to the presumed edge of the map to the east is proportionately greater than to the west.

Certainly, though, it is drawn more spaciously that any of our other large Burchard maps, giving the mapmaker free range for his pictorial skills. Towns are shown as realistic groups of buildings with towers and spires, though fantasy creeps in with the tall pyramids at Gadara and Pella. At Ramoth-Gilead, shown with a city wall, we

67. Positions of towns on the Pierpont Morgan map and the grid maps. The thick lines show where towns are drawn on the Pierpont Morgan map, while the dots mark the squares where there are towns on the grid maps (from the copy of Sanudo's index in BL, Add. MS 27376*, ff. 139v-141v). There is little correspondence between the two maps.

11 I am most grateful to Prof. Patrick Gautier Dalché and to Dr Maurice Vandermaesen for their very positive help in my investigation of these fragments.

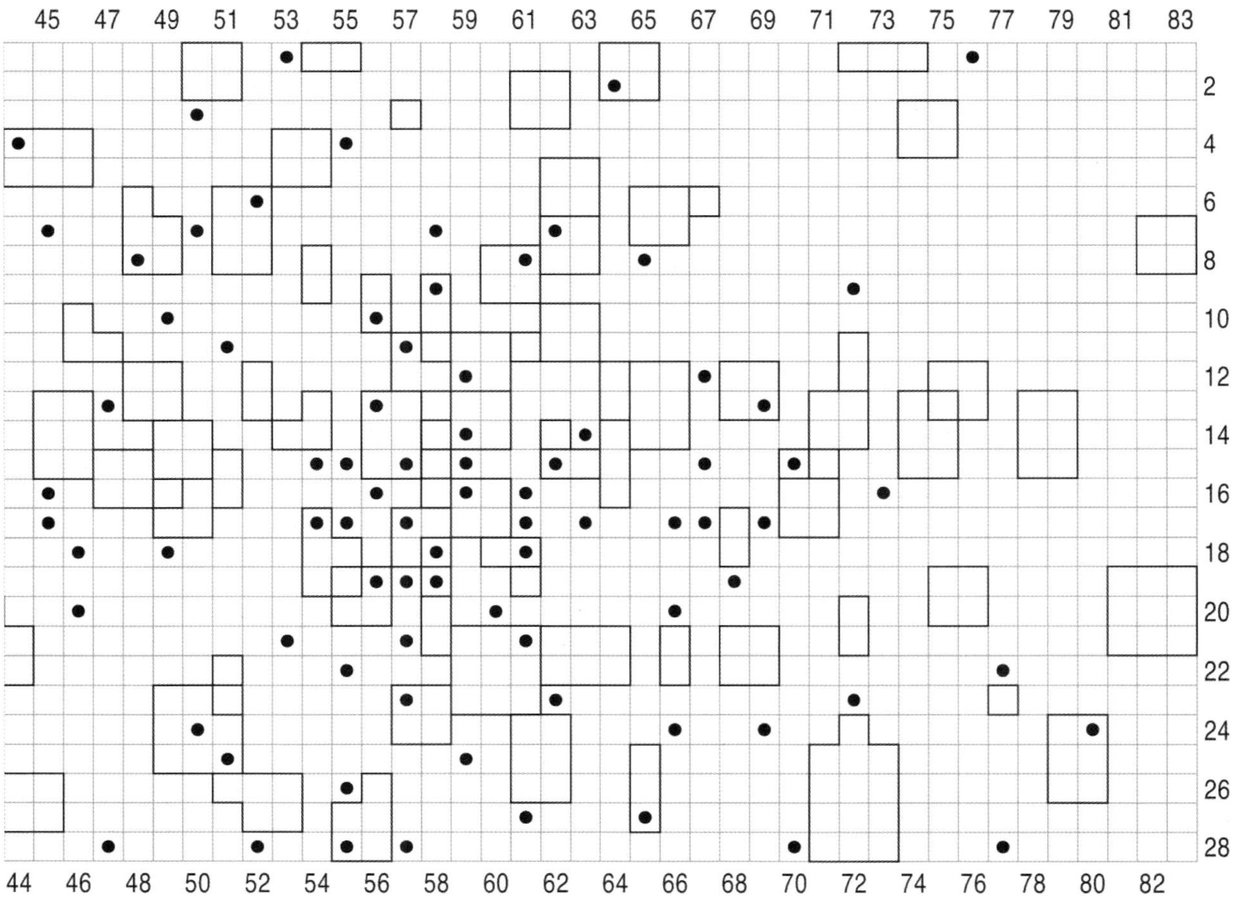

recognise the tall square towers of contemporary Italy. The towns are placed not on conventional hillocks but on carefully drawn hills, so that many have the appearance of Italian hill towns. On the eastern fragment there are many trees, again realistically drawn, and four boats on the sea of Galilee, all unparalleled on our earlier large Burchard maps. As on the other maps towns may be drawn sideways or, as at Tiberias, upside down so as to face the lake. Rivers are drawn with a single line, so the tribal boundaries can be distinguished only by being drawn in red instead of brown.

Only brown and red ink are used for the map; there are no coloured features. Van de Walle notes that some of the more important wording is in red. This is difficult to distinguish on the images but it probably includes *tiberi*[*as siue*] *genes*[*aret*] on the western fragment, and on the eastern *Basan* (i.e. Bashan) and the first words, *Mare*

68. The eastern Bruges fragment, 14th century; east at the top. Below is the south end of the sea of Galilee with four boats; the central town on a broad hill is Ramoth-Gilead and above it, to the right, is Macheronta. Bruges, Archives de l'état, Chartes avec numéro bleu 11458, 1st fragment.

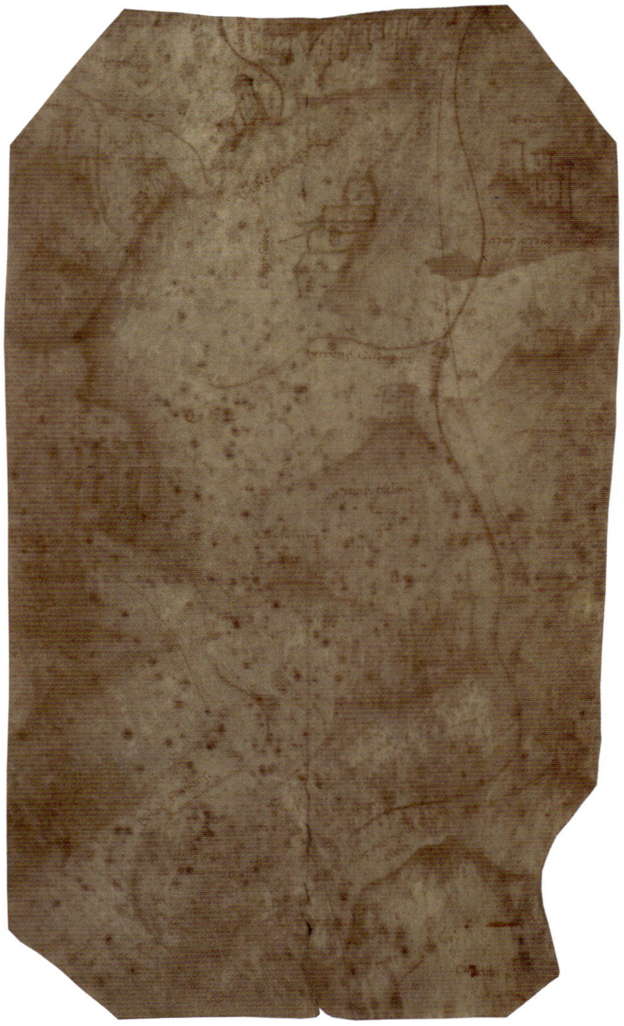

69. The western Bruges fragment, 14th century; east at the top. At the top the buildings drawn upside down are Tiberias, on the west shore of the (cut off) sea of Galilee; Mount Tabor is in the centre and Mount Carmel in the lower right corner. Bruges, Archives de l'état, Chartes avec numéro bleu 11458, 2nd fragment.

and *Hic*, of two inscriptions, bottom right; one gives the name of the sea of Galilee, the other, mostly lost, probably noted the start of the Jordan valley. The red may have been used for regional and other names that were not simply names of settlements. Each of the four boats on the sea of Galilee has a red cross on its sail, and Van de Walle convincingly identifies two red eye-like shapes as paragraph marks, each meant to introduce the name of a tribe that has not in fact been put in. The tribes not named were presumably Zabulun on the western fragment, Gad on the eastern.

There are two indications that the map stemmed from the same original exemplar as the Archivio di Stato and Pierpont Morgan maps, not from the rather better source of the grid maps: Jerash is named *Gorza*, and though we have no more than the first two letters of the name of Caymon, these are *Ch*, recalling the *Chaimborat* of our earliest map. Very faint traces on the images of what may be one or two horizontal and vertical lines perhaps come from a grid; if so, it was drawn lightly, simply for copying, and may well have been largely erased. While the Bruges fragments give us no more than a glimpse of a splendid map now lost, they add significantly to our impression that quite a number of copies of the large Burchard map were made in the course of the fourteenth century, and that their failure to survive is simply typical of the fate of so many large sheet maps in all periods.

Happily, the map in the Douce manuscripts in the Bodleian Library at Oxford has survived intact, indeed in very good condition.[12] Apart from minor damage it is complete and the colours are bright and clear. A crack has developed along the join of the first and second of its three membranes and there are tiny holes in the parchment elsewhere, as well as some fraying along the bottom edge. Also along the bottom edge is a series of larger holes, perhaps the work of a mouse, that show that the map was at some point kept rolled up, and it is indeed called a roll (*rotulus*) in the 1840 catalogue of the Douce collection.[13] This will account for further fraying at the right-hand end, presumably the outside of the roll. It is now kept flat, fully visible but held in a mount in a glazed wooden frame; the dorse is not accessible. The fraying of the bottom and right-hand edges, and some of the lacunae on the first membrane, have been repaired with parchment, and the whole map appears to be mounted on linen backing. We may suppose that the map was repaired soon after it reached the Bodleian Library, which happened after the death of Francis Douce in 1834. Douce was a collector of books, manuscripts and other antiquities; he had been Keeper of Manuscripts at the British Museum from 1807 to 1811 but bequeathed his manuscripts to the Bodleian Library.[14] How or when he acquired the map we do not know; from its handwriting and style we can say that it was made in the late fourteenth century in England, which is interesting evidence of the wide distribution – and thus probably many copies – of the large Burchard map.[15]

It is unquestionably the best-known of the large Burchard maps. In 1857 the Roxburghe Club printed for its members an edition of William Wey's account of his travels; a Fellow successively of Exeter College, Oxford, and of Eton College, Wey had made a pilgrimage to Compostela in 1456 and two journeys to Palestine, in 1457-8 and 1462.[16] The Club followed this in 1867 by issuing a coloured lithographic facsimile of the Douce map, entitling it 'Map of the Holy Land, illustrating

12 I am grateful to Dr Pnina Arad for helpful discussion of the Douce map.

13 *Catalogue of the printed books and manuscripts bequeathed by Francis Douce, Esq. to the Bodleian Library* (Oxford, 1840), p. 68.

14 C. Hurst in *ODNB*, XVI, pp. 594-5.

15 Otto Pächt and J.J.G. Alexander, *Illuminated manuscripts in the Bodleian Library Oxford* (3 vols, Oxford, 1966-73), III, p. 67.

16 *Itineraries of William Wey*.

the intineraries of William Wey'.[17] In his text, Wey describes 'my map of the Holy Land' and the Roxburghe Club's editor, George Williams, seeing many similarities, had persuaded himself that this was the Douce map and that its connection with Wey 'may be said to have been demonstrated';[18] this was hardly correct for, as we shall see, there are many dissimilarities as well. However, Williams's assessment, along with the map's publication as a companion to Wey's text, has led some to suppose that this was indeed Wey's map and that he not only owned it but was its author – though Williams himself had pointed out that it is not in Wey's hand.[19] In fact it was drawn well before Wey's time and there is no reason to connect it with Wey at all.[20]

The Roxburghe Club's facsimile has been reproduced a number of times, in whole or in part, and in 1904 Reinhold Röhricht published a line drawing based on it, but it was not until 1994 that a photograph of the original was published, in colour but much reduced.[21] This has mattered less than one might suppose as the drawing and lettering of the facsimile are extremely accurate, and while the colours diverge from those of the original they are for the most part consistently applied; thus the map's light brown has mostly been printed in red and the pink used on the second and third membranes appears as yellow.

The map's proportions differ from the other large Burchard maps. It is as long from north to south as we suppose the map of the Bruges fragments was, but it is much narrower from east to west. At the north end the first 20 cm – but no more – are taken up with a list of distances between places, of which two-thirds (down to the sixteen miles from Mount Carmel to Caesarea Palestina) comprise the successive distances embedded in the mid-twelfth-century account of the Holy Land by Rorgo Fretellus.[22] The last sixteen entries (starting with the ten miles from Hebron to the Dead Sea) are from an unidentified source or sources; unlike the earlier distances, all in miles, two of them are in leagues, and while some, such as the four miles from Jericho to the Jordan, are found in many texts, others, such as the eight miles from Cana to Chorazin, are unusual.

From the list at the north end, south as far as Jerusalem, the map is very like the Archivio di Stato and Pierpont Morgan maps; everything is compressed from east to west, but the same places are named and we find the same inscriptions. Despite the compression, the map even places on land the long inscription that lies off the coast on the Archivio di Stato map. The wording of the inscriptions differs only slightly from the Archivio di Stato map; thus the two that are compared with Burchard's text in chapter 9 (above, p. 104) here read:

> Hunc fontem iosephus appellat phialam Marcus dalmanucha Matheus Magedam Saracena modan et est velud ortus iordanis phalee enim in hunc fontem mille in dan recipiuntur subteraneo meatu ducte
>
> In hac silo sunt tabernaculum et archa domini que nunc dicitur ad sanctum samuelem

We see small changes in transmission but no more.

On the Archivio di Stato map the distance from Jerusalem to the map's southern edge is more than one-third of its whole length from north to south, and on the Pierpont Morgan map it is more than a quarter. On the Douce map, however, it is less than one-sixth, so that at this end the map has been compressed from north to

17 *Map of the Holy Land.*

18 *Itineraries of William Wey*, p. xvi; however, when this was reprinted in introducing the facsimile in 1867 the anonymous editor was a little more cautious (*Map of the Holy Land*, p. iv).

19 *Itineraries of William Wey*, p. xvi.

20 As pointed out by Falconer Madan in 1897 (*A summary catalogue of western manuscripts in the Bodleian Library at Oxford*, ed. by Falconer Madan et al. (7 vols, Oxford, 1895-1953), IV, pp. 617-18).

21 Nebenzahl, *Maps of the Holy Land*, pp. 50-2; Reinhold Röhricht, 'Die Palästinakarte des William Wey', *Zeitschrift des deutschen Palästina-Vereins*, 27 (1904), plates VIII, IX; *A continental shelf: books across Europe from Ptolemy to Don Quixote: an exhibition to mark the re-opening of the Bodleian Exhibition Room* (Oxford, 1994), pp. 28-9. In 1965 the map was exhibited in London at the Victoria and Albert Museum (*World of the Bible: centenary exhibition of the Palestine Exploration Fund* (London, 1965), p. 48).

22 *Rorgo Fretellus*, ed. Boeren. There are three versions of the text, but the differences between them do not affect the notes of distances.

south even more than from east to west. This was possible only because towns and other features are sparser in this area than further north, but there is considerable distortion all the same. Most notably, Hebron has been moved far south, so that the Brook Besor flows right through it. Even so, space has been found for the inscription identifying Beersheba as the southern limit of the Land of Promise, but we have only the first few words of the account of the desert of Kadesh and the third of the longer inscriptions at this end of the map, on the Brook Besor itself, is wholly missing.

There is no grid on the map and no trace of erased grid lines. Eight of the twelve tribes are named but without tribal boundaries and perhaps the tribes of Asher and Benjamin are each named three times simply to show where their areas extended; Naphtali is named twice, but in one case probably by mistake for Zebulun, and though the half-tribes of Manasseh are named as being one on each side of the Jordan, both are placed west of the river. The tribes of Dan, Judah and Simeon are not named, probably for want of space at the south end of the map. Rivers and lakes, drawn with red or brown outlines, are coloured blue and the sea and mountain areas, all with brown outlines, are green. Towns are marked with buildings, drawn with rather more enthusiasm than finesse; they are more elaborate than those on the Archivio di Stato map and, while they do not reach the height of fantasy achieved on the Pierpont Morgan map, their colouring is more varied. The town symbols are unrelated to those on either of the other two maps or on grid map H; even Job's tomb is drawn without a spire and there is no more than a gesture towards the triangular shape of Heber the Kenite's tent. In the Dead Sea are the ghostly submerged shapes of the four destroyed cities of the plain.

A single hand wrote all the map's inscriptions as well as the list of distances. However – and this is perhaps the map's oddest feature – each of its three membranes was drawn and coloured by a different draftsman. This appears impressionistically in a simple overall view of the map, but the differences between each membrane can be precisely defined. On the first membrane, for example, but not on the second or third, anonymous figures appear over the battlements or at the windows of some of the buildings, and there are rabbits on some of the hillocks below; again, on the third membrane, but not on the first or second, several buildings have chimneys that emit puffs of red fire. The most obvious of these differences are listed in the appendix to this chapter, pp. 139-40; they may mean that the workshop had to produce the map in haste, or they may hint at some form of mass production, on however small a scale.

At least the outlines of the map's drawn features were in place before the inscriptions were entered, neatly and precisely, in the spaces left between them; the colouring within the outlines may have been done afterwards. Certainly the outlines were drawn and the inscriptions written before the three membranes were stuck together; the overlap at each join is not cut straight but goes around drawn features and (at one point between the first and second membranes) an inscription **[70]**. But rivers and mountains inevitably cross from one membrane to the next and here we see how the few inexactitudes have been adjusted once the membranes had been put together. Above the names of six towns on the second membrane are tiny arabic numerals in a greyer shade of ink than the map's inscriptions;[23] they cannot be much, if at all, later than the map itself.

23 Tiberias 10, Nazareth 12, Nain 14, Mount Tabor 16, Jenin 22, Sebaste 23.

While the Pierpont Morgan map and the Bruges fragments repeat mistakes that we found on the Archivio di Stato map, the Douce map seems to stem from the rather better copy of the original map that lay behind the grid maps. The tribes of Gad and Reuben are the right way round and, looking at place-names garbled on the other large Burchard maps, here we have, more correctly, *Geth* and *Cainmon*. Again, *Goraca* and *Assur* are not far removed from the grid maps' *Gerasa* and *Arsur* and Joshua's burial-place, the Vulgate's Thamnathsare, is *Tampnathsare*, as on grid map B. It may well be significant that Jerusalem is here marked with a magnificent building instead of the cross in a circle of the other two maps. It looks as if we may distinguish two lines of copies stemming from the original large Burchard map; the grid maps and the Douce map belonged to the more correct one, the other large maps and probably the Bruges fragments to the other.

The manuscript of Wey's travels is in the Bodleian Library, Oxford,[24] and on the flyleaf is a note, in English, of items he bequeathed to the house of Bonhommes at Edington in Wiltshire that he joined in about 1467 – he died there in 1476.[25] Among them were 'also a mappa mundy also a mappa of the holylond wyth Ierusalem in the myddys also ij levys of parchement on wyth the tempyl of Ierusalem another wyth the holy movnte of olyvete'.[26] To the accounts of his journeys Wey appended (1) a list of 'everything on the map of the Holy Land' (*omnia in mappa terre sancte*), giving apparently all its inscriptions and moving broadly from north to south; (2) a list arranged in alphabetical order of initial letter of all the place-names 'on my map of the Holy Land' (*in mappa mea de terra sancta*); (3) a list of 'distances between places in the Holy Land' (*Distancia locorum in terra sancta*).[27] He does not say that this list of distances came from his – or any – map, but his editor was probably right to assume that it was there, like the list of distances on the Douce map. It is in fact a very similar list; some of the distances differ, but these are figures that were often miscopied, and though the order of the entries departs from that of Fretellus this was clearly the source for most of them, just as on the Douce map. We may envisage a version of the Douce map on which the list of distances had undergone changes in successive copyings. The description of Jerusalem as in the middle of the map need not be taken literally.

However, the lists of names, first topographical and then alphabetical, are far longer than any list that could be compiled from the Douce map. Certainly there is a clear relationship: the forms of the names and the notes of events correspond to those on the Douce map. But at two points the lists diverge sharply from the surviving map. They include local names in and beside Jerusalem: the places where the disciples slept the first and the second time, the house of Pilate, the gate and the street through which Christ passed to his Passion and many more. It is not just that these do not appear on the Douce map; there could not possibly be room for them. And in the south the lists do not stop at the Brook Besor and the adjacent deserts as the map does, but extend far into Egypt with a further fifty names, the Red Sea, Cairo and the Nile among them. We might suppose that Wey's map was a much enhanced version of the large Burchard map, extended further to the south and with space around Jerusalem for local features. Alternatively – and this seems more likely – Wey took the names for Jerusalem and Egypt from other sources and inserted them into his list at the

24 MS Bodley 565.

25 Henry Summerson in *ODNB*, LVIII, pp. 333-4; an account of the house of Bonhommes is in the *Victoria history of Wiltshire* (1953-), III, pp. 320-1.

26 MS Bodley 565, f. 2v; *Itineraries of William Wey*, pp. xxviii-xxix.

27 MS Bodley 565, ff. 83v-90v; *Itineraries of William Wey*, pp. 128-40.

28 *Itineraries of William Wey*, p. 129, has 'Caynar' but it is in fact 'Caynon' in MS Bodley 565, f. 84r.

29 However, Thamnathsare is here 'Tamnascere' (MS Bodley 565, f. 84v; *Itineraries of William Wey*, p. 130, reads it as 'Taumascere').

appropriate points; his leaves of parchment with the Temple and the mount of Olives might well have provided the names for Jerusalem. For what it is worth, the names on the lists seem closer to the Douce map than to the other large Burchard maps: the tribe of Gad is named before Reuben, and *Geth*, *Caynon*,[28] *Goroca* and *Dora vel Assur* all recall the forms we have seen on the Douce map.[29]

70. Boundary of two membranes of the Douce map. The Douce map was drawn before its three membranes were glued together, so the parchment was cut in a curved line following the map's features; this appears clearly here where the first and second membranes meet.

Our three large Burchard maps, the Bruges fragments and the map described by William Wey, as well as the grid maps, all derive from a single map that must have been one of the most remarkable achievements of medieval mapmaking. It was contemporary with the Hereford and Ebstorf world maps, but it was a completely different kind of cartography. Whether the mapmaker worked from Burchard's text or from personal knowledge or from both – and it is not impossible that he was himself Burchard – it was the direct graphic representation of a region, not a compilation of earlier sources constructed in the study. We may or may not be right in seeing in our surviving maps two lines of descent from the prototype; the reality may have been a good deal more complicated. We clearly have only a proportion of the grid maps that were actually produced, perhaps – at a guess – between a quarter and a half. But a map in a book has a far greater chance of survival than a sheet map; despite its size and complexity the large Burchard map may well have been not only one of the most impressive maps produced in medieval Europe but also one of the most often copied.

APPENDIX

SOME OF THE DIFFERENCES IN THE ARTWORK BETWEEN THE THREE MEMBRANES OF THE DOUCE MAP

1 Olive-green is used on mem. 1, bright green on mm. 2 and 3; light brown is used on all three membranes, but dark brown and pink only on mm. 2 and 3; dark red-brown is used only on mem. 3.

2 Much green is used in colouring buildings on mem. 1, less on mem. 2 and very little on mem. 3.

3 Many battlements are coloured red on mem. 3, few on mem. 2 and none on mem. 1.

4 The rivers are outlined in red on mm. 1 and 2, but in brown on mem. 3.

5 Anonymous figures appear over the battlements or at the windows of some of the buildings on mem. 1, and there are rabbits on some of the hillocks below, but not on mm. 2 and 3.

6 Several buildings on mem. 3, but not on mm. 1 and 2, have chimneys that emit puffs of red fire: Jerusalem, Hebron, *Coronaim*, Rama, Acharon, Gath.

7 The name of each tribe is written in a red ornamental frame with green trefoils on mem. 1, in a red ornamental frame without trefoils on mem. 2, without a frame on mem. 3 (except a faint brown frame-line around 'Tribus iuda').

8 The indentations of the mountain chains are most intricate on mem. 1, less so on mem. 3 and still less on mem. 2, where they are merely wavy lines.

9 The green hillock below each town symbol is precisely outlined on mem. 1, less so on mem. 3 and still less on mem. 2, where in some cases the outline at the foot is omitted altogether.

THE SMALL BURCHARD MAP
mid-14th century

12

Florence, Biblioteca Medicea Laurenziana, ms Plut. 76.56, ff. 97v-98r **[71]**. Parchment, 33 × 45 cm. Brown, black and red ink.

IN LOOKING AT THE MEDIEVAL MAPS of Palestine we have seen how copyists have altered, moved or reversed the names on their exemplars, or have introduced serious distortions to fit a map into the available space, but none has achieved the chaos of the only copy we have of the small Burchard map, in a manuscript volume where it accompanies a copy of the full version of Burchard's text. At first glance it seems a reasonably acceptable map, with east at the top. We see that the river Jordan flows from the mountains of Lebanon through the water of Maron and the sea of Galilee to the Dead Sea; that Jerusalem is appropriately placed in relation to this; that the Red Sea borders the map on the right with the Egyptian lands beyond; and that the Mediterranean coast curves round on the left to face north instead of west – thus fitting conveniently into the double page on which it is drawn. But then we find that the twin sources of the Jordan lie not far from the coast, but far from the rest of the river, that Damascus ('Here the blessed Paul was converted') lies even nearer the coast and that a whole block of names from the heart of the country, Sebaste and Nablus among them, has been moved west to the coast between Lydda and Caesarea Palestina. What we have must be a travesty of what the original mapmaker intended – but it is a travesty full of interest.

The double page with the map is at the centre of an eight-leaf gathering that contains the copy of Burchard's work; the text extends on to the two pages at the back of the map and one might suppose that the map was drawn to illustrate the text on this opening, left blank on purpose. In fact this is probably the opposite of what happened. The parchment of the double page is thicker and coarser than the other leaves in the gathering,[1] and it has been trimmed to make it the same size, cutting thorough words at top and bottom. It seems most likely that the map, which is in a different hand from the text, was drawn first, and was used as a peg from which to hang a copy of Burchard's text. Careful but straightforward calculation would establish the number of leaves required before and after the map to contain the whole work.

The gathering with Burchard's work and the map is the fourteenth in a volume occupied mostly – on the first twelve gatherings – by a copy of Boethius, *Consolation of philosophy* (*De consolatione philosophiae*) with the commentary by the English Dominican Nicholas Trevet. This is in a different hand from either the map or Burchard's text, and the text by Cicero on the fifteenth and final gathering is in a different hand again. However, on the thirteenth gathering are excerpts from works

1 That the parchment is coarser explains why the text on the back of the map is written on lines 4.0 mm apart, instead of the 3.5 mm of the rest of the text.

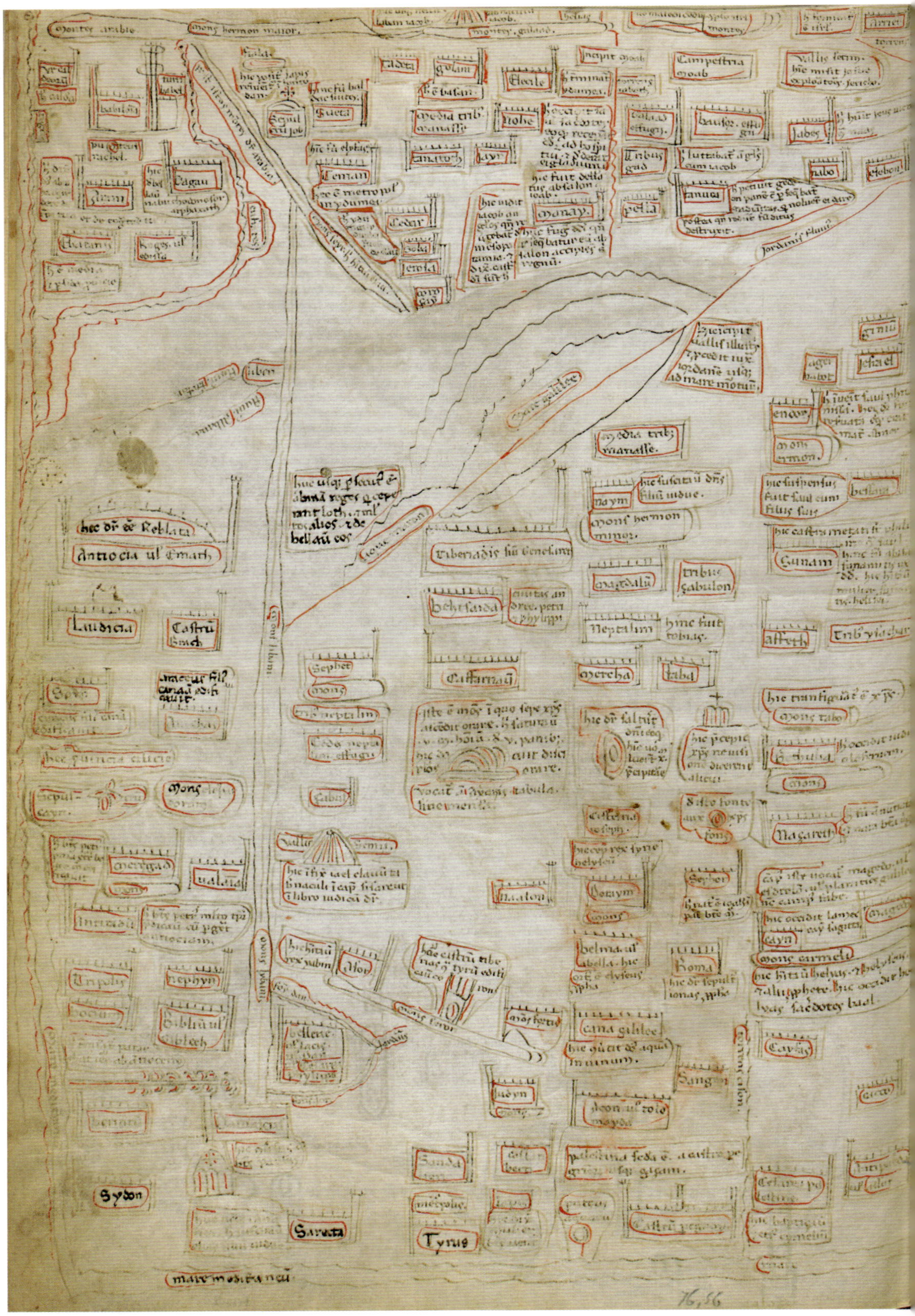
Sydon
Tyrus

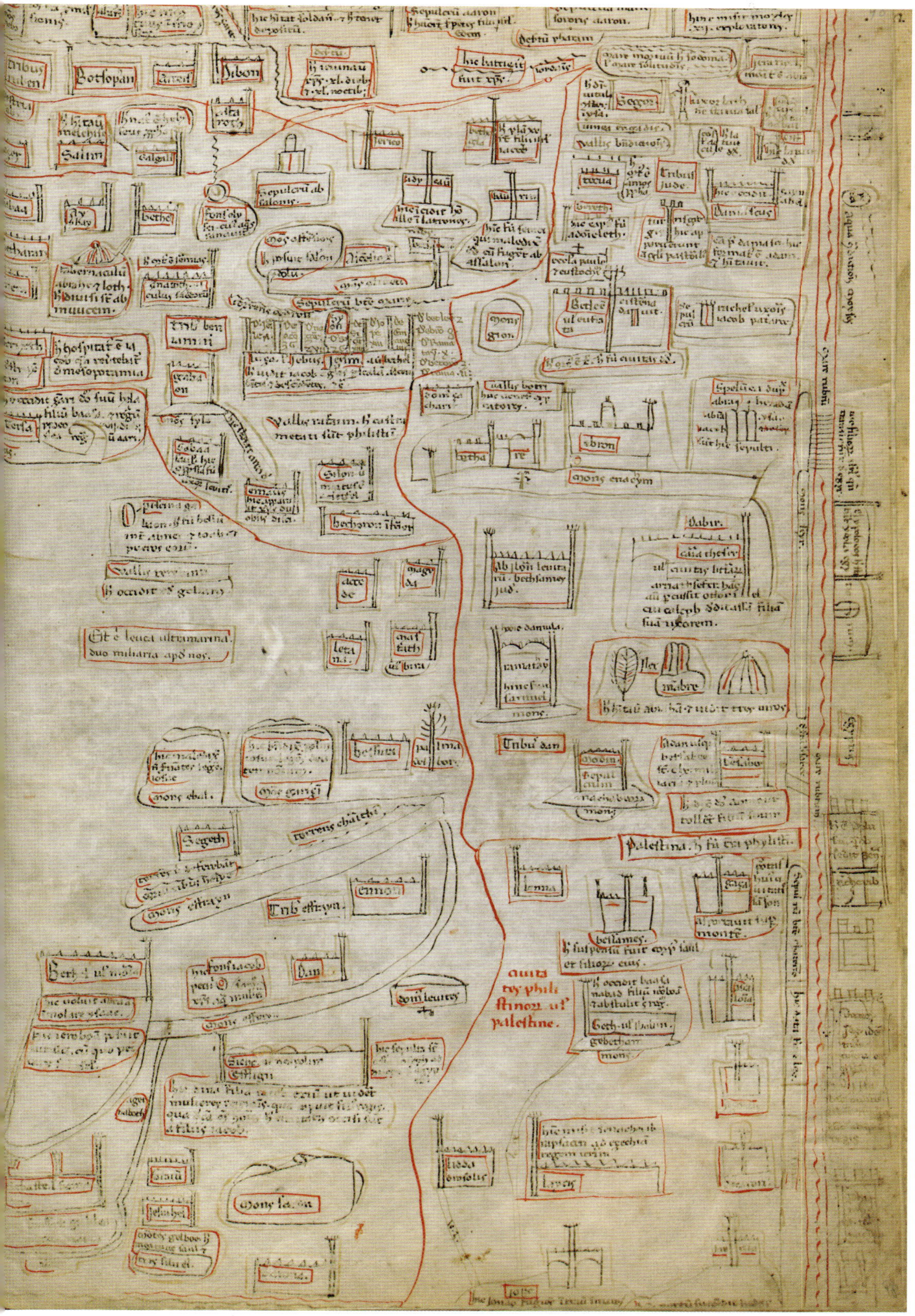

71. The small Burchard map, mid-14th century; east at the top. The Mediterranean coast runs along the foot of the map and up the left-hand side; along the right-hand side is the Red Sea. The semi-circular sea of Galilee, upper left, is far larger than the oblong Dead Sea in the top right corner. Florence, Biblioteca Medicea Laurenziana, ms Plut. 76.56, ff. 97v-98r.

by Eusebius (in Latin) and Jerome and these are in the same hand as Burchard's text that follows them on the fourteenth. Probably these two gatherings were written together and joined with the other unrelated works in a volume made up and bound at a later date; they are written (apart from the bifolium with the map) on thinner parchment than the rest of the volume.

The map of Palestine is not the only adornment of Burchard's text. At the foot of the page before the map is a stylised plan of Jerusalem, apparently unrelated to any others that have yet been reproduced and discussed **[72]**.[2] And throughout the text there are words or phrases in the margins – some 140 in all – that serve as a guide to the contents, such as names of towns and other sites. On the last three pages, containing Burchard's accounts of the twelve tribes, the natural history and the religions of the Holy Land, each of these marginal notes is enclosed in a simple frame-line, but in the rest of the text this frame is elaborated to reflect the wording it encloses. For many towns it takes the form of a building with a battlemented tower, but some towns and other places have more complex designs; thus for the site of the miracles of loaves and fishes there is a fish in a rectangle **[73]**.[3] For the Dead Sea there is a map-like thumbnail sketch, an oblong of wavy lines, which is very like the Dead Sea on the map,[4] and the general style of these frame-lines is close to the way places are marked there; probably all were inspired by the map, and in some cases we can detect a direct relationship. The turreted triangle that marks Acre in the margin echoes the triangle that marks the well of living waters, nearby on the map, though it may simply recall Burchard's description of Acre, 'having a triangular shape, like a shield' (*triangulam habens formam, ut clypeus*).[5] For the chapel at Sarepta where Elijah revived the widow's son, the text's margin has a dome-like shape with crosses that recalls what on the map looks like a triple lancet with rose window.[6]

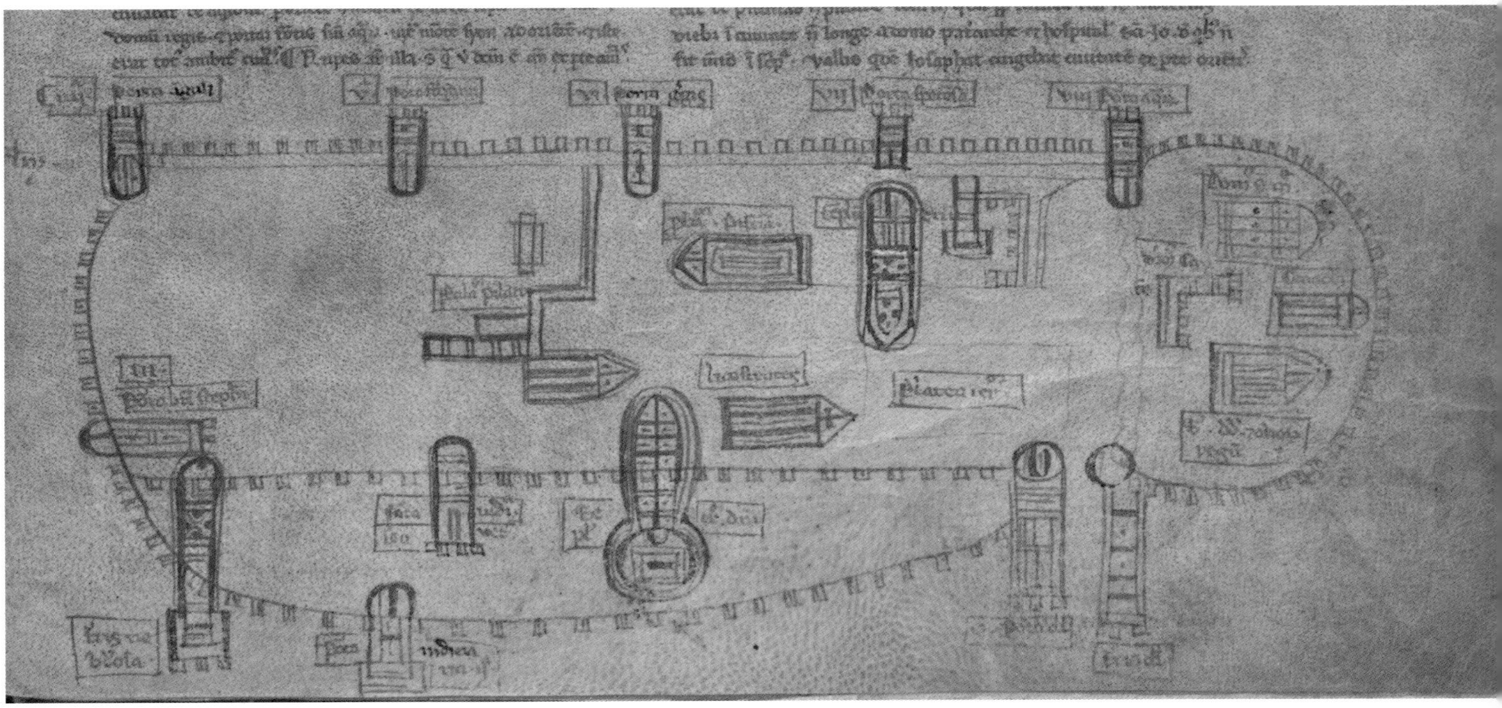

72. Plan of Jerusalem illustrating Burchard's text, mid-14th century; south at the top. Drawn as a further illustration to Burchard's description of the Holy Land on the page before the map, the plan is in the hand of the writer of the text, not of the mapmaker. Florence, Biblioteca Medicea Laurenziana, ms Plut. 76.56, f. 97r.

2 A redrawn and typeset reproduction is in Röhricht, 'Marino Sanudo', plate 8. The plan is reproduced and described in Vilnay, *Holy Land*, p. 18, and, with fuller discussion, in Hebrew, by Milka Levy in *The history of Jerusalem: crusaders and Ayyubids (1099-1250)*, ed. by J. Prawer and H. Ben-Shammai (Jerusalem, 1991), pp. 483-6, a reference I owe to the kindness of Dr Hanna Vorholt.

3 f. 95r.

4 f. 96v.

5 f. 94r; *Peregrinatores*, ed. Laurent, p. 23.

6 f. 94v; I Kings (III Kings in the Vulgate) 17.17-24.

Most towns on the map are marked with a building, usually battlemented, with a single slender tower, all crudely drawn, but some are more elaborate, like the two towers and a dome on the building at Hebron. Some sites, as we have seen, have a specially appropriate symbol, but always sketchy, never well drawn – we have here no artist at work. Red is used to frame many names, and for the outline and waves of the Red Sea, and there is a solitary inscription in red, in the lower right corner, 'The cities of the Philistines or of Palestine' (*ciuitates philistinorum uel palestine*). The other red lines on the map probably mark rivers, though the river Kishon is shown as a wavy brown line and some double lines in the south-west may be either rivers or roads. Mountain areas are outlined in brown and a series of horizontal lines marks the Israelites' passage of the Red Sea. All the tribes are named except Asher and Simeon, but without their boundaries. The map is in a single hand but a few notes and names on the left side are in a black ink distinct from the brown of the rest.

The full version of Burchard's text is the source of most, though not all, of the names and notes on the map. Except in its north-west quarter, some come from other sources, among them, as Reinhold Röhricht shows, the *Liber de locis*, Jerome's translation of the *Onomasticon* of Eusebius.[7] Among the notes taken from Burchard it is easier than on the large Burchard map to find more or less exact quotations rather than paraphrases. Thus at Tortosa (*Anteradum*) the map reads:

> beatus petrus multo tempore predicauit cum pergeret antiociam

and Burchard's text is:

> beatus Petrus multo tempore predicauit, cum Antiochiam pergeret.[8]

Again, at Kadesh-barnea the map has:

> Hinc misit moyses .xii. exploratores

and Burchard:

> De quo Moyses misit xii exploratores.[9]

At Jerusalem a list of distances (in leagues) to seven other towns is taken from the list entered there by Burchard, and to four further places from other points in his text;[10] six of the numbers are entered on the map in arabic, not roman, numerals. That

7 Röhricht, 'Marino Sanudo', pp. 104-5.
8 *Peregrinatores*, ed. Laurent, p. 30.
9 Ibid., p. 80.
10 Ibid., pp. 56, 63, 77.

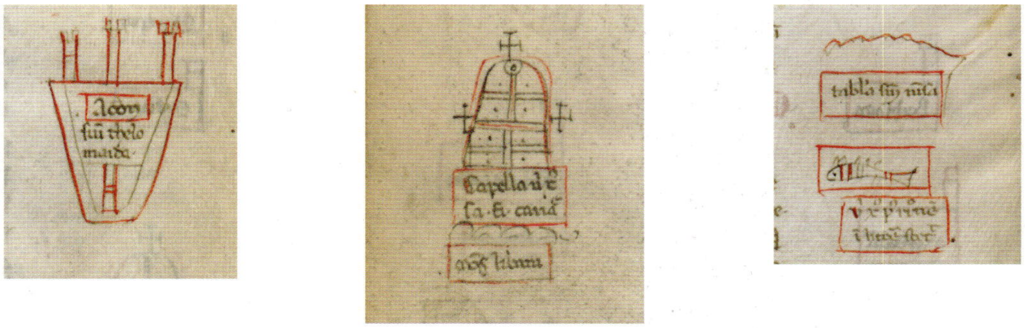

73. Thumbnail drawings illustrating Burchard's text, mid-14th century. Examples of the marginal drawings in the copy of the text that accompanies the small Burchard map: Acre (left), the chapel at Sarepta where Elijah revived the widow's son (wrongly titled where Christ cured the nobleman's son at Cana) and Mount Lebanon (centre), the sites in Galilee of the sermon on the mount, the miracles of loaves and fishes and Christ's appearance after his resurrection (right). Florence, Biblioteca Medicea Laurenziana, ms Plut. 76.56, ff. 94r,v, 95r.

not all the distances exactly correspond to those in the edited text of Burchard can be confidently explained as mistakes in copying. These are far from being the only copyists' errors on the map; not only have many names been badly misplaced, many have been badly miscopied. Thus what is most likely Thamnathsare appears as *Coracum*, while the alternative name given to the Dead Sea, the sea of Solitude, *mare solitudinis*, should probably be the sea of Salts, *mare salinarum*.[11]

By what processes of miscopying – or perhaps of mismapping in the first place – the map attained its extraordinary state of confusion is difficult even to guess. There is little point in trying to do so until we know more of the hundred-odd manuscripts known to contain copies of Burchard's work. Here we have seen the text added to an existing map, not the map to the text, but the two will have been seen as companion pieces and it is not so much possible as likely that other copies of the map will be found in the many unexplored copies of the text. Even where manuscripts have been listed in a library's printed catalogue, such a map would not necessarily be mentioned. We should keep in mind the possibility that our small Burchard map derives, albeit in travestied form, from the map that Burchard himself sent to his friend at Magdeburg, along with the short version of his text. More than for any of the other maps discussed here, all that can be offered for the small Burchard map is a first impression, a preliminary assessment.

The only published account of the small Burchard map is a short description by Röhricht in 1898; with it he produced a typeset reproduction of the map's wording, omitting all lines and symbols – understandably, for they would have made the reproduction impossibly complicated.[12] He followed his description with accounts of two late-fifteenth-century maps that he associated with Burchard's text, and to these we may add a third, of the early sixteenth century. These three maps are related to each other; they are effectively versions of the same map.[13] Whether they derive ultimately from the same archetype as what we have called the small Burchard map is much less certain; it cannot be formally demonstrated and we can say only that it is not beyond the bounds of possibility. However, two of the three accompany copies of Burchard's work, the third a closely related text, and there are some points of contact with the earlier map. Crucially, they are the same shape: they are all, like the small Burchard map, on a double-page opening of a book. On the left side of all four maps the Mediterranean coast turns to run eastward, then either turns to run north again or continues eastward, but unrealistically bordering the region of Damascus and beyond. The Red Sea bounds all four maps on the south, with either the Israelites' crossing or Mount Sinai towards its east end.

The first of the three further maps is in the *Rudimentum noviciorum* (*Guide for beginners*), a massive history of the world that was printed at Lübeck in 1475 by Lucas Brandis. Included in it is the first printed edition of Burchard's work; the full version of his text occupies ff. 162r-188r of the book, and at the beginning is the map, on ff. 162v-163r **[74]**.[14] The two pages of this opening are not a bifolium and the map is printed from two separate woodblocks, one for each page, with wording in inserted type. Throughout it is a picture map. The land is covered with craggy or rounded hills seen from the west, and on nearly all there is a name. Mostly this is the name of a town, and the hill has on top a realistic view of towers and other buildings, but some

11 Röhricht, 'Marino Sanudo', p. 105, lists these and many other errors in place-names.

12 Ibid., pp. 104-5, plate 7.

13 Note, for instance, the (bizarre) relative positions of Acre and Caesarea Palestina on both the *Rudimentum noviciorum* map and the manuscript map, and the positions of Damascus, Mount Sinai and Egypt in three corners of the two printed maps.

14 The map is described and reproduced by Tony Campbell, *The earliest printed maps 1472-1500* (London, 1987), pp. 146-8.

are the names of mountains and other sites, as *Jor et dan* at the sources of the Jordan and the salt statue that was Lot's wife. With some of the names are references to chapters of the Bible; thus *Nobe i re xxj* refers to David's meeting with Ahimelech at Nob, I Samuel 21 (I Kings 21 in the Vulgate). There are, however, no further notes, whether from Burchard or from any other source. Here and there are trees and roads with figures walking along them, and there are a few biblical scenes: Moses on Mount Sinai, the baptism of Christ, the crowned or helmeted heads of Pharaoh and his army drowning in the Red Sea. At the centre is Jerusalem, large and detailed, with outside it the crucified Christ at Calvary, and Acre is also drawn as a large walled city. In the sea are ships and around the edge of the map are eight men's heads, unnamed but blowing hard to represent the winds. Two later versions of the map were produced, each from new woodblocks, the first with wording cut on the block, the second with inserted type; both illustrated *La mer des hystoires*, a French version of the *Rudimentum*, the first published at Paris in 1488 and again about 1503, the other at Lyon in 1491, with successive editions down to at least 1555.[15]

The second printed map is quite different, though it is likely that it too came from the press of Lucas Brandis at Lübeck in about 1478. It is in a book entitled *Prologus Arminensis in mappam Terresancte, templi domini, ac sancte ciuitatis Hierusalem*, 'Herman's introduction to the map of the Holy Land, of the Lord's Temple and of the holy city Jerusalem'; the author is not further identified and neither the printer nor the date is named.[16] The book is in four sections, first an account of the Temple in Jerusalem, then an account of the whole city, then (the longest section) an account of all Palestine based largely on Burchard and finally a discussion of the chronology of the two books of Maccabees; at the end is an alphabetical index. Each of the first three sections has at the start a plan or map consisting simply of typeset names and other wording; there are no lines, symbols or pictures. Each plan or map serves as an index to its section, giving the number or letter of the chapter in the book after the relevant wording, but on all three there are further names without a reference. The plans of the Temple and of Jerusalem occupy each a single page, but the map of Palestine covers a full opening, ff. [11]v-[12]r (neither leaves nor quires are numbered) **[75]**; the first chapter of the account of Palestine begins on f. [10]v and the last, the 150th, is on f. [25]v. There are no biblical notes or references on the map, but three of the tribes are named: Manasseh, Benjamin and Judah. At the foot, notes explain the system of references and the difficulty of placing names on the exact spot, given, particularly, the number of names in Galilee. Around the edge of the map twelve winds are named.

The manuscript map is unlike either of the printed maps. It is on a double-page opening at the end of an early-sixteenth-century copy of Burchard's work that was one of the two manuscripts that J.C.M. Laurent used as the basis for his edition in 1864; Laurent mentions the map in his note on the manuscript, but does not describe it **[76]**.[17] It consists of wording and lines, with no more than a hint of symbols or pictures: groups of feather-like signs at Mount Hermon and in the desert between Jerusalem and Jericho. On the left-hand page Acre, and on the right-hand page Jerusalem, have each a small circle drawn around the name, with eight lines radiating from it – just possibly a now meaningless vestige of a diagram showing the divisions

15 Ibid., pp. 149-51.

16 A facsimile edition with introduction by W.A. Neumann was published by the Société de l'Orient latin as *Prologus Arminensis in mappam Terrae Sanctae Lubicae ante annum MCDLXXX impressus* (Geneva, 1885); Neumann there suggests that the author was Hermann von Sina, a Lübeck Dominican. Another suggestion is that it was written by Johannes von dem Berge, an Augustinian canon (*Bruchstücke zur Kenntnis der Lübecker Erstdrucke von 1464 bis 1524* (Lübeck, 1903), pp. 2-3).

17 Hamburg, Staats- und Universitätsbibliothek, Codex geographicus 59, pp. 70-1; *Peregrinatores*, ed. Laurent, p. 6. I am most grateful to Dr Hans-Walter Stork of the library's manuscripts department for identifying the manuscript and for providing me with images.

74. Map in *Rudimentum noviciorum*, illustrating the first printed edition of Burchard's text, published at Lübeck, 1475; east at the top. On the left the coast turns northwards, placing Acre at the end of a deep bay; along the right side of the map is the Red Sea, with Mount Sinai in the top right corner. *Rudimentum noviciorum*, ff. 162v-163r.

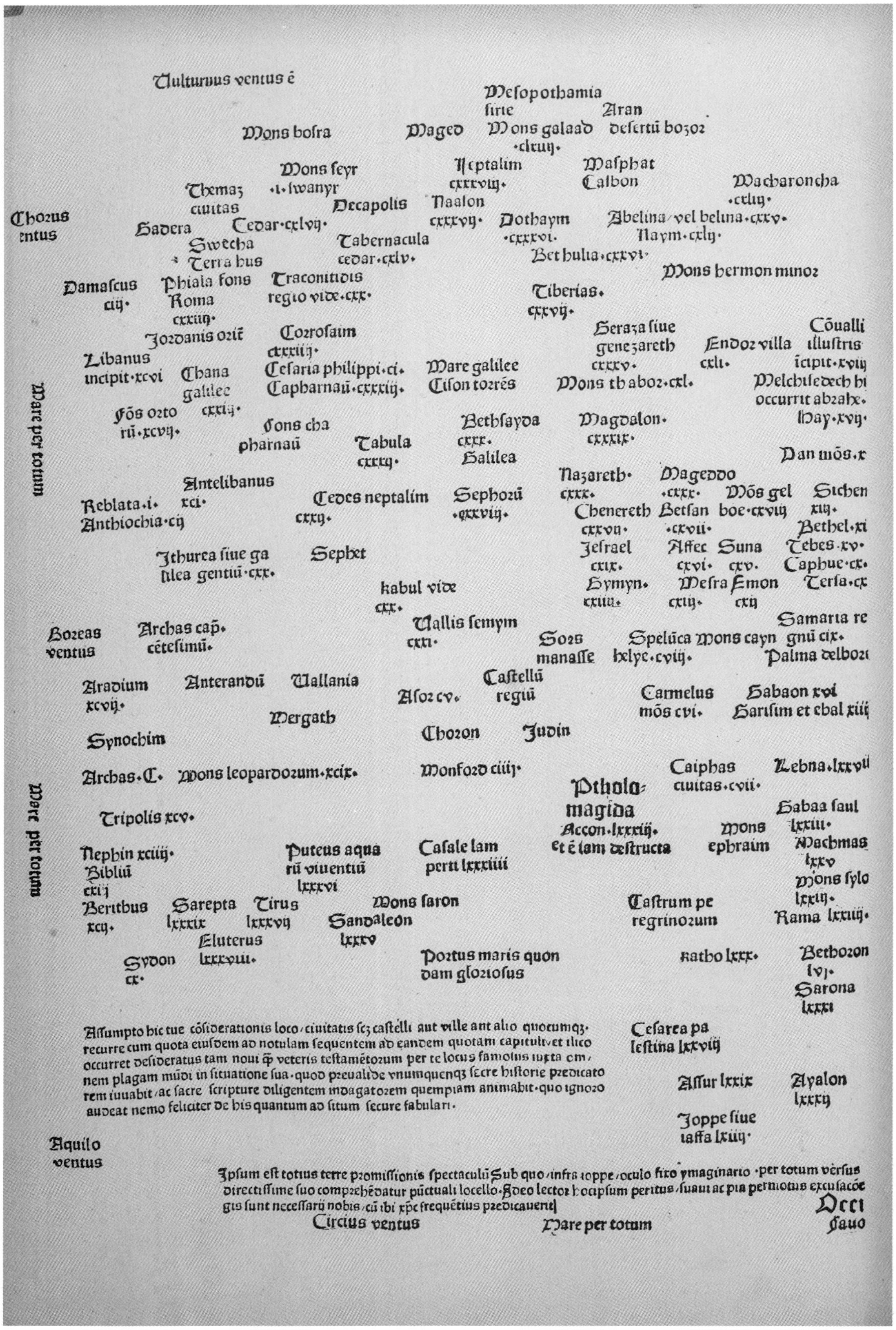
Vulturnus ventus e
Mesopothamia sirie
Aran
Mons bosra
Mageo
Mons galaad
desertu bozoz
Mons seyr
Neptalim
Masphat
Themaz ciuitas
Decapolis
Naalon
Macharoncha
Chorus entus
Sadera
Cedar
Dothaym
Abelina vel belina
Tabernacula cedar
Naym
Swecha
Terra hus
Bethulia
Mons hermon minor
Damascus
Phiala fons
Roma
Traconitidis regio vide
Tiberias
Jordanis orit
Corrosaim
Geraza siue genezareth
Endor villa
Libanus incipit
Chana galilee
Cesaria philippi
Capharnaum
Mare galilee
Cison torres
Mons thabor
Melchisedech
Fons orto
Fons cha pharnau
Tabula
Bethsayda
Galilea
Magdalon
Day
Dan mos
Mare per totum
Antelibanus
Reblata
Anthiochia
Cedes neptalim
Sephoru
Nazareth
Mageddo
Mos gel boe
Sichen
Chenereth
Betsan
Bethel
Ithurea siue galilea gentiu
Sephet
Jesrael
Affec
Suna
Tebes
Caphue
Kabul vide
Symyn
Mesra
Emon
Tersa
Vallis semym
Samaria regnu
Boreas ventus
Archas cap. cetesimu
Sozs manasse
Speluca
Mons cayn
Palma delbor
Aradium
Anterandu
Vallania
Castellu regiu
Afoz
Carmelus mos
Gabaon
Garisim et ebal
Mergath
Synochim
Chozon
Judin
Archas
Mons leopardozum
Monfozd
Caiphas ciuitas
Lebna
Ptholomaida
Accon
et e iam destructa
Tripolis
Gabaa saul
Mons ephraim
Machmas
Nephin
Bibliu
Puteus aqua ru viuentiu
Casale lampertu
Mons sylo
Beritbus
Sarepta
Tirus
Mons saron
Sandaleon
Castrum peregrinozum
Rama
Eluterus
Sydon
Portus maris quondam gloziosus
Rathbo
Bethozon
Sarona
Cesarea palestina
Assur
Ayalon
Joppe siue iaffa
Aquilo ventus
Circius ventus
Mare per totum
Occi fauo

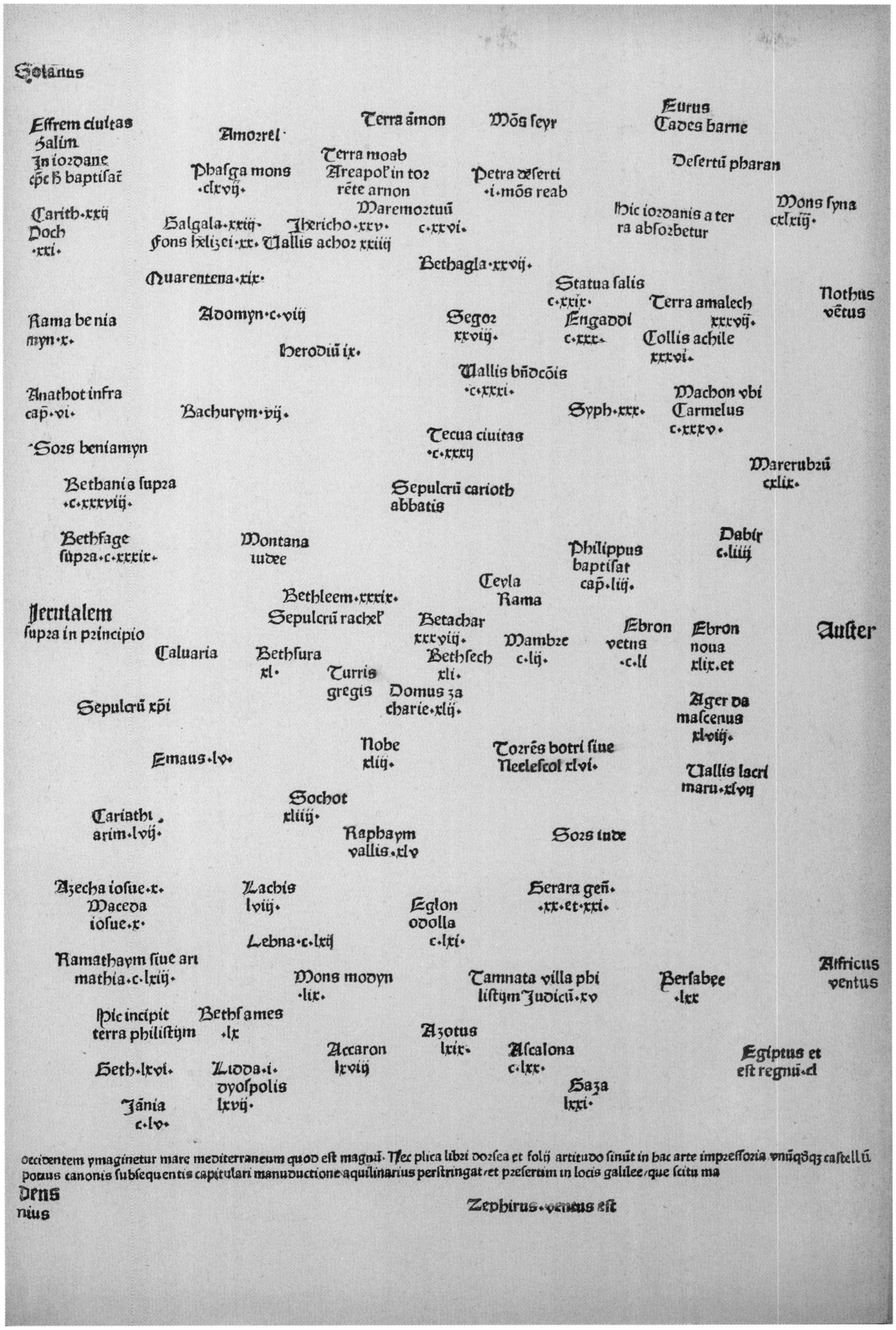

75. Map in *Prologus Arminensis*, published probably at Lübeck, 1478; east at the top. The map consists entirely of type-set place-names, with figures referring to the chapters of the book's account of Palestine; around the edge are the names of the winds. *Prologus Arminensis*, ff. [11]v-[12]r.

of Burchard's text based on Acre.[18] While the names of the coastal towns show that the coast runs east and west between Caesarea Palestina and Acre, the line of the Mediterranean runs north and south regardless, right along the foot of the map, leaving the map's north-west quarter in a kind of limbo, neither land nor sea. Wavy lines mark the river Kishon, south of Acre, and the Jordan with the sea of Galilee and the Dead Sea, and there are three estuaries along the southern part of the coast; mountains, however, are marked only by their names. Most of the names on the map

18 There is such a diagram on p. 13 of the same manuscript [50].

are of towns or other biblical sites, and many are accompanied by heavily abbreviated notes. Some, as on the map in the *Rudimentum noviciorum*, are references to the Bible: *3 R 18* beside the name of the Kishon means chapter 18 in the first book of Kings (third book of Kings in the Vulgate), which tells how Elijah killed the priests of Baal there. Others give distances from place to place, taken from Burchard; thus Bethlehem is two leagues from Jerusalem, Cana four leagues from Acre.[19] A few give other information, also taken from Burchard: Ascalon is beside the sea and was a city

19 *Peregrinatores*, ed. Laurent, pp. 78, 44.

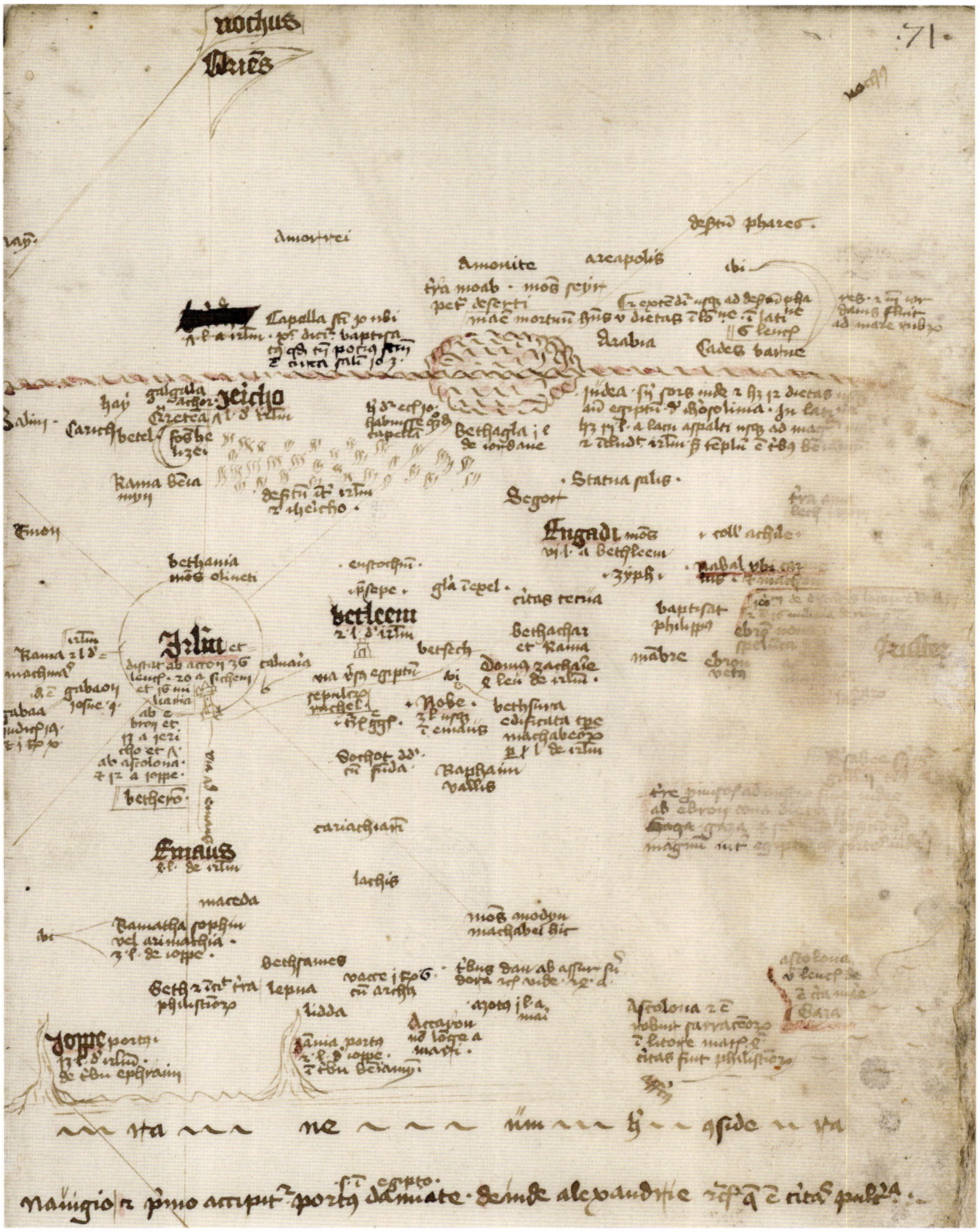

76. Map illustrating Burchard's text, 16th century; east at the top. A circle with radiating lines marks Jerusalem on the right-hand page, Acre on the left. The few notes give information taken from Burchard. Hamburg, Staats- und Universitätsbibliothek, Codex geographicus 59, pp. 70-1.

of the Philistines, the Trachonitis region is called the plain of Lebanon in the book of Joshua.[20] These do not correspond to the quotations from Burchard on our small Burchard map. Seven tribes are named or mentioned west of the Jordan, and east of the river a note names the two tribes and a half that were settled there. Around the map the four cardinal points are named and two winds, *Affricus* (the south-west wind) on the west side and, confusingly, *Nothus* (the south wind) on the east.

That these three maps, while related, are so different from each other suggests that many more derived from their original archetype – this at a period when drawing and using maps was becoming much more usual than a hundred years or so earlier. This archetype, along with its derivatives, may or may not be part of the story of the small Burchard map. But that is a story still to be told.

20 Ibid., pp. 85, 33.

CONCLUSION

13

IN THE LAST EIGHT CHAPTERS we have looked in detail at twenty-three artefacts that represent eight different maps of Palestine: the Ashburnham Libri map, the three Tournai maps, Matthew Paris's Oxford and Acre maps and the large and small Burchard maps. Throughout it must be obvious how much is owed to earlier work; what is done here is to build on this or to carry it forward. This earlier work was done mostly more than a century ago and mostly by German-speaking scholars; only for the maps of Matthew Paris and the grid maps has later work made a significant contribution, and of the others only the third Tournai map has been noticed at all in work on medieval cartography. This is extraordinary, for all these maps are of great interest and one, the large Burchard map, is among the most remarkable maps that medieval Europe produced.[1] While some of this earlier work has had to be revised – notably the ascription of the third Tournai map to Jerome – it was careful, learned and thorough, providing a firm basis for further research and analysis. Without the work of, above all, Reinhold Röhricht and Konrad Miller this book could not have been written.

Underlying this further research and analysis is the recognition that in only two of these eight regional maps of Palestine do we have what the original mapmaker meant us to see; these two are the third Tournai map and Matthew Paris's Acre map. For the others we can reconstruct what the mapmaker had in mind only by varying processes of stripping away excrescences, deciphering erased wording, comparing different derivatives and so on. This, of course, is our loss; it would be wonderful to have still the original large Burchard map or the possibly sumptuous production that Matthew Paris rapidly and sketchily copied in his Oxford map. But in looking at the maps as we have them it is immensely helpful to realise that what may seem egregious errors of the mapmaker are no more than evidence of careless copying.

While these maps certainly add to our understanding of medieval cartography, they conform to a well-established pattern. All the maps down to Matthew Paris's Acre map in the mid-thirteenth century come from England or north-east France. It was here that the encyclopaedic world map developed in the course of the twelfth and thirteenth centuries and it is unsurprising that the starting-point of every one of these maps of Palestine seems to be an extract from a world map, an extract that might be reinforced by information drawn from accounts of the Holy Land and other written sources. The genesis of our small Burchard map is elusive, but when we look at the large Burchard map and its grid-map derivatives we move into a different cartographic world, a world centred on Italy, the world of the portolan charts and of the realistic town plans of Jerusalem, Acre and Antioch that accompany some of the grid maps, where the map is created by direct observation more or less accurately recorded.

1 I am among those guilty of this neglect. In my contribution, on local and regional maps, to *History of cartography*, pp. 464-501, I make no more than a passing reference to the large Burchard map (on p. 477) and in Harvey, *Medieval maps*, I mention only Matthew Paris's maps, the grid maps and the Tournai map of Asia.

This change reflects changes in who drew the maps and who used them. The Tournai maps and Matthew Paris's Acre map were produced by monks for their fellow-monks and the same is probably true of the Ashburnham Libri map. Whatever their immediate context, they would be seen as learned adjuncts to the learned works that were required reading in the monasteries, works that directly or indirectly were focused on the Bible and on the lands that the maps displayed. All will have been for devout and knowledgeable perusal and reflection by those dedicated to the spiritual life. The map that Matthew Paris copied in his Oxford map may too have hung in a monastery; or it may have hung in a noble or royal household, again for reflection on the events of scripture but also as a constant reminder of the call of the Holy Land for pilgrimage and, above all, for crusade.

We might well say the same of the large Burchard map, and indeed its derivative grid maps were made and distributed with a work that had just this end in view. But we should also see it in the light of Burchard's own work, which he wrote so that those unable to go as pilgrims to the Holy Land could make the pilgrimage in imagination, achieving at every site the same emotional and spiritual experience as if they were actually there. The map could serve the same purpose with its helpful notes (from Burchard's text) explaining the biblical associations of so many places. It will have been designed not just for monks or other religious, but for layfolk as well, and is as likely to have originated in a secular workshop as in a monastic scriptorium. If significant numbers of the large Burchard map were in fact produced until at least the end of the fourteenth century, it is in these workshops that they will have been made and we should see them in the context of the growth of personal piety and devotion in late-medieval Europe.

The present book can be no more than a staging-post in our understanding of these maps, for further work will certainly reveal much more. We should look to work not only on the maps themselves but on written texts as well. We have seen how Patrick Gautier Dalché's discovery of the descriptive text at Naples throws a completely new light on the third Tournai map of Palestine and on the accompanying map of Asia. Investigations leading to more precise knowledge of the amorphous complexes of texts used in the Ashburnham Libri map and Matthew Paris's Oxford map could tell us significantly more about both. Above all we need a definitive edition of Burchard's text in both its full and its shortened versions, based on thorough exploration of the many surviving manuscripts. As a literary work of high quality it deserves no less; that it would also tell us more about the large Burchard map, a great deal more about the small Burchard map, is a secondary consideration but all the same an important one.

GENERAL INDEX

INDEX OF MANUSCRIPTS

Listed here are the pages in chapters 5 to 12 where these manuscripts are described and discussed